Game Theory Unbound

Revolutionize Your Thinking and Learn How to Win in Life and Business. Think Like a Strategist, Predict Outcomes, Play Smarter, and Navigate Life's Games with Confidence

by
Thinkleap

Table of Contents

Welcome Onboard and How to Get Your Bonus Content 5

Part 1 - Foundations of Game Theory 6
Section 1: Getting Started 6
Section 2: Basics of Game Theory 9
Section 3: Understanding Rational Decision Making 23

Part 2 - Core Game Theory Concepts 41
Section 4: Nash Equilibrium and Strategic Stability 41
Section 5: The Prisoner's Dilemma & Other Example "Games" 46
Section 6: Zero-Sum and Non-Zero-Sum Games 59

Part 3 - Advanced Strategies and Applications 66
Section 7: Repeated and Sequential Games 66
Section 8: Mixed Strategies and Randomness 73
Section 9: Bayesian Games and Information Asymmetry 80
Section 10: Evolutionary Game Theory 86
Section 11: Mechanism Design - Shaping Decisions and Systems 96

Part 4 - Real Life and Concrete Applications of Game Theory 109
Section 12: Game theory in personal life and everyday decisions. 109
Section 13: Economics and Market Analysis Through the Lens of Game Theory 110
Section 14: Political Science and Game Theory 112
Section 15: Military Strategy and Game Theory 114
Section 16: Computer Science and Networking 115
Section 17: Exploring Game Theory in Evolutionary Biology 117
Section 18: Corporate Strategy and Negotiations 118
Section 19: Game Theory in Business 120
Section 20: Exploring Game Theory in Behavioral Economics 122
Section 21: Game Theory Applications in Finance 124
Section 22: Game Theory in Law and Regulation 125
Section 23: Game Theory in Social Network Analysis 127
Section 24: Game Theory in Public Policy and Resource Allocation 128
Section 25: Game Theory in Environmental Strategy 129
Section 26: Game Theory in Healthcare and Epidemiology 130
Section 27: Game Theory in Sports 131

The Key Insights and Lessons from Game Theory to Remember 134

Conclusion 135

Thank You! 137

Get Your Bonus Content 138

References 139

Disclaimer & Legal Notice

Please be aware that the information provided in this document is intended solely for educational and entertainment purposes. While every effort has been made to ensure the information is accurate, current, reliable, and complete, no guarantees are made regarding its correctness or completeness. The reader should understand that the author does not provide legal, financial, medical, or any other form of professional advice. Readers are advised to consult a licensed professional before applying any techniques discussed herein.

By accessing this document, the reader agrees that the author shall not be liable for any direct, indirect, incidental, or consequential losses that may arise from the use of information contained in this document. This includes losses attributed to errors, omissions, or inaccuracies.

Copyright © 2024

All Rights Reserved. Reproducing, distributing, or transmitting any part of this document by any means—including but not limited to photocopying, recording, or other electronic or mechanical methods—is strictly prohibited without the express written permission of the publisher. Exceptions are granted for brief quotations used in critical reviews and certain noncommercial uses as permitted under copyright law.

Welcome Onboard and How to Get Your Bonus Content

Welcome to this exciting journey! We're so glad you've chosen to embark on this adventure. Writing this book has been an incredible journey, filled with hard work, extensive research, and numerous interviews with experts in the field. This labor of love was driven by a passion to share knowledge that can enrich everyone's lives, both personally and professionally. Our goal is to make complex concepts accessible and engaging, hoping to inspire and empower readers with tools that can transform their thinking and approach to challenges.

Before diving in, here are a few key steps you can take:

1) Get Your Bonus Content, jump to the section at the end of the book titled 'Get Your Bonus Content' to further amplify your learning and get into action faster!

2) Share with the World, why not sharing how thrilled you are about starting this new book. Your enthusiasm can inspire others to join in and discover the insights that await them.

Taking a moment to share your thoughts can make a huge difference, and it won't take more than a minute of your time. Here's how you can help spread the word:

Option 1: Capture a short video showing off your new book. Let others see the excitement in your eyes!

Option 2: Not keen on being in front of the camera? No problem at all! Take one or more photos of the book and add a brief comment about your initial impressions.

Option 3: If you prefer to keep things super simple, even a few written words from you would be absolutely wonderful.

Thank you for taking the time to share your excitement. P.S. Please note that this is entirely optional.

Scan the QR code below to leave your review:

Part 1 - Foundations of Game Theory

Welcome to the foundational segment of our journey through the world of game theory. This part of the book is designed to build a robust understanding of game theory from the ground up. Whether you're new to the concept or looking to deepen your existing knowledge, these chapters will guide you through the basic principles and applications of game theory in a variety of contexts.

We begin by defining what game theory is and exploring its broad scope within decision-making. This lays the groundwork by introducing the essential components of any game—players, strategies, payoffs, and information sets. We'll discuss how these elements interact to shape the decisions made by individuals and groups in different scenarios. This discussion will extend into understanding the purpose and use of game theory across various fields, illustrating its versatility and impact.

We'll delve into the different types of games, such as cooperative versus non-cooperative games, static versus dynamic games, and games with complete versus incomplete information. Each type offers unique insights and challenges, which we'll explore through practical examples. Additionally, this chapter applies game theory to everyday life, helping you understand how it can influence personal decisions, career paths, business strategies, sports tactics, social interactions, and even environmental management.

After we shift our focus to the concept of rationality in decision-making. We'll define what it means to make a "rational" decision and discuss the criteria for rationality across different scenarios. This chapter introduces the Expected Utility Theory, which provides a framework for understanding how individuals assess risk and make choices that maximize their satisfaction.

However, rational decision-making has its limitations and challenges, which we will acknowledge and examine. Recognizing these limitations will prepare you to better apply game theory in realistic settings, where pure rationality often gives way to other psychological and social factors. We will also explore how game theory applies to personal finance, career planning, business negotiations, government policy-making, legal strategies, and environmental management, providing a comprehensive view of its utility and significance in our daily lives and societal functions.

This part of the book is more than just an academic exploration; it's a practical guide designed to equip you with the tools to view the world differently—through the lens of strategic interaction and rational decision-making. By the end of this part, you'll have a solid foundation in game theory, enabling you to start thinking like a strategist in both personal and professional arenas.

Section 1: Getting Started

Have you ever paused, in the middle of a decision, wondering not only about your choice but also how it might influence, or be influenced by, the choices of others? This book ventures deep into those very reflections, transforming the abstract elements of game theory into a practical toolkit for your everyday life. Whether it's negotiating a better salary, deciding the best strategy for your commute, or navigating complex family dynamics, game theory offers profound insights that can help steer these decisions with precision and insight.

Imagine that each day is a series of games, where the rules are not laid out in a rulebook but are instead shaped by the interactions and choices of those involved. These are not just games of chance; they are strategic games where your success hinges not only on your actions but also on anticipating the moves of others. From boardroom tactics to social engagements, game theory provides a lens through which we can view and effectively influence our interactions.

Game theory started as a niche scholarly concept, mainly confined to economics and mathematics, but has since permeated various fields including psychology, politics, and everyday decision-making. It equips us to analyze and strategize effectively, taking into account the actions and reactions of others. Understanding these dynamics is crucial, as it empowers you to navigate life's complexities with a strategist's mindset.

This book is designed for anyone intrigued by the challenge of making better decisions. No prior expertise in mathematics or economics is needed. All you need is a willingness to think strategically about the decisions you face. We'll cover the basics of recognizing patterns in decisions, predicting outcomes, and crafting strategies that account for various scenarios you might encounter.

Through engaging stories and real-world applications, we'll demonstrate how the principles of game theory apply not only to monumental decisions but also to everyday choices. Each chapter will unravel these concepts further, supplemented by practical exercises and examples to connect theory with the realities of daily life.

As we progress, you'll find that you're already participating in numerous "games." Whether you're deliberating on career moves, managing investments, or planning family activities, each decision is a move in these strategic games. By the end of this book, you'll possess a robust set of tools that will enhance your ability to assess situations, make informed decisions, and foresee the moves of others more adeptly. Game theory isn't just about understanding games; it's about transforming your approach to life's challenges and opportunities.

How to Navigate and Make the Most of This Book

The primary purpose of this book is to demystify game theory and present it in a way that is accessible and applicable to everyday life. While game theory may seem like a domain reserved for economists and strategists, this book aims to show that everyone can benefit from understanding its principles. Whether you're making decisions about your career, your finances, or your personal relationships, game theory provides a set of tools that can help you consider these decisions more strategically and effectively.

This book is designed not only to introduce the concepts of game theory but also to empower you to use these concepts to enhance your decision-making processes. By the end of this book, you should be able to apply game theory principles to a variety of real-life situations, helping you to achieve better outcomes and understand the dynamics of interactions with others.

You will notice that certain concepts and examples are intentionally revisited and discussed in multiple contexts. This little repetition is by design, aimed at enhancing comprehension (as we say "skills grow from repetition") and also allowing us to explore these ideas from various angles. By examining the same principles through different lenses—whether in economic strategies, social interactions, or complex decision-making scenarios—we can deepen our understanding and appreciation of their broad applicability and nuanced implications. This approach helps to solidify the foundational knowledge of game theory while illustrating how its versatile tools can be adapted to address diverse challenges and opportunities across a wide range of disciplines.

Making the Most of This Book

The book is divided into several parts, each building upon the knowledge of the previous. It starts with foundational concepts and gradually moves into more complex theories and their applications. This structured approach helps you grasp the essentials before diving into more intricate applications.

Below, you'll find a guide designed to help you engage with and utilize the contents of this book to improve your decision-making and strategic planning skills.

1. Understand the Basic Concepts: Before diving into the applications, it's crucial that you grasp the fundamental concepts of game theory. Spend time with the initial chapters that outline these principles, ensuring you understand terms like "Nash Equilibrium," "zero-sum games," and "mixed strategies." These concepts are the building blocks for everything that follows.

2. Reflect on the Examples: Each concept introduced in the book is accompanied by one or more examples that illustrate how it can be applied in real-world scenarios. Pause at these examples and consider how the outcomes might change with different decisions or under different circumstances. This reflection will help you think critically about applying game theory to your own situations.

3. Use the Reflection Questions: Reflection questions prompt you to think deeply about how game theory relates to your experiences and decisions. These questions are designed to bridge the gap between theory and practice, helping you to see the relevance of game theory in everyday life.

5. Apply the Concepts: Beyond just reading and understanding, attempt to apply the concepts of game theory to your daily decisions. Whether it's a negotiation, a planning decision, or managing a conflict, try to think about these situations in terms of game theory. This practical application will make the concepts much more tangible and useful.

Practical Application Tips

1. Daily Decisions: Start small by applying game theory to daily decisions, such as how you negotiate tasks at work or manage your time. Ask yourself what the potential outcomes are, consider what others involved might prefer, and strategize based on this analysis.

2. Conflict Resolution: Use game theory to approach conflicts more constructively, whether in personal relationships or professional settings. Identify the players, their preferences, and how different resolutions will affect those preferences. This can lead to more empathetic and effective solutions.

3. Planning and Negotiation: When planning an event or negotiating an agreement, think several steps ahead, just as you would in a chess game. Anticipate responses and plan your moves accordingly, aiming for outcomes that offer mutual benefits to strengthen relationships and achieve better results.

4. Strategic Career Decisions: Apply game theory when considering career moves, such as whether to pursue a promotion or switch industries. Analyze the payoffs of different scenarios, considering not only your potential gains but also the risks and benefits from the perspective of others involved, like your employer or family.

Long-Term Engagement

Keep the Book Handy: As you become more comfortable with these concepts, keep the book as a reference to revisit concepts and examples. Game theory is a dense subject, and its applications can become more nuanced as your understanding deepens.

Continuous Learning: Game theory is a dynamic field, and its applications are continually evolving. Stay engaged by reading current articles, watching relevant talks, and even participating in forums or discussion groups that focus on the practical applications of game theory.

Teach Others: One of the best ways to deepen your understanding of a subject is to explain it to someone else. Share your knowledge of game theory with friends or colleagues. Discussing the concepts can provide new insights and enhance your grasp of the material.

Section 2: Basics of Game Theory

What is Game Theory?

At its simplest, game theory is the mathematical study of strategy and decision-making where multiple players make choices that affect each other. Game theory involves multiple participants, each with their own interests and objectives. It explores the principles of rational decision-making across humans, animals, and computational systems. It's a framework that helps you, and others involved, determine optimal strategies when your outcomes depend not just on your own decisions but also on the decisions of others.

The formal development of game theory began in the 20th century, but its concepts have been applied, knowingly or unknowingly, throughout human history. The foundational work was laid out by John von Neumann and Oskar Morgenstern in their 1944 book, "Theory of Games and Economic Behavior." This groundbreaking work proposed that economic behavior could be studied as strategic games.

In the decades since, game theory has influenced countless fields, including economics, biology, computer science, politics, and psychology. Its ability to provide theoretical underpinning for understanding complex scenarios where the outcome for each participant depends critically on the actions of others has made it an indispensable tool in these disciplines.

Imagine you're at a bidding auction for a piece of art that you and several others are eager to buy. Each bidder's final outcome, whether they go home with the art or not, and how much they pay if they do, depends significantly on what others are willing to bid. Game theory provides the tools to analyze such situations—helping you decide whether to bid, how much to bid, or whether to bow out, based on predicting others' behavior.

The significance of game theory extends beyond pure economics or business. It applies to any situation involving individuals or entities with different goals and the need to make decisions based on predicting others' actions. This could be as simple as choosing routes during your commute to avoid traffic (anticipating the choices of other drivers) or as complex as negotiating an international trade deal.

Game theory's real power lies in its ability to provide structured insights into the potential outcomes of various strategies, helping you make better-informed decisions that consider not only your own objectives but also those of others who might impact your results. The scope of game theory is vast and interdisciplinary, impacting numerous fields such as:

- **Economics, Finance and Business**: Understanding competitive strategies, market equilibria, auction designs, and more.
- **Politics**: Analyzing voting systems, coalition formations, and legislative tactics.
- **Social Sciences**: Examining social norms, group dynamics, and network effects.
- **Biology**: Studying evolutionary strategies and population dynamics.
- **Computer Science**: Developing algorithms for machine learning, network security, and optimization problems.

In each of these areas, game theory offers a set of analytical tools to model complex interactions and predict the equilibrium state where no participant can benefit by changing their strategy unilaterally.

By using game theory, you can transform your approach to decision-making from reactive to strategic. Instead of making decisions in isolation, you begin to see the web of interconnections, anticipating how others might respond to your moves. This strategic foresight can be crucial in both personal and professional contexts. Game Theory can be instrumental for:

- **Personal Decisions**: Choosing when to buy a house, deciding on marriage, or planning retirement savings are all scenarios where understanding the decisions of others can be beneficial.
- **Professional Decisions**: From competitive business decisions to internal company negotiations, knowing how to model these scenarios using game theory can provide a significant advantage.

Components of a Game: Players, strategies, payoffs, and information sets

Ever find yourself pondering your next move in life, not just for yourself but also considering how others might react? Whether you're negotiating for a raise, deciding the best route to work, or figuring out the best approach to resolve a family issue, game theory offers insights that help you navigate these choices with clarity and confidence. Here will explore the four critical components of game theory—players, strategies, payoffs, and information sets—and show you how they can illuminate the hidden strategies behind daily interactions, empowering you to lead a more thoughtful and strategic life.

Players: The Decision-Makers in Game Theory

In any game theory model, players are the individuals or entities making decisions that affect the outcomes of the game. Players can be real or hypothetical and range from single individuals in a personal decision-making scenario to large groups or even nations in complex geopolitical strategies. Characteristics of Players:

- **Rationality:** Players are assumed to be rational, meaning they always make decisions that they believe will provide them with the greatest benefit. This assumption is crucial because it allows us to predict how players might behave under different circumstances.
- **Knowledge:** The information available to players can vary widely. Some games assume perfect knowledge, where all players know everything about the game, including the strategies and payoffs of other players. More commonly, games involve imperfect information, where players do not have full knowledge and must make decisions under uncertainty.
- **Number of Players:** The dynamics of a game can change significantly depending on the number of participants. Games can be as simple as two-player interactions or as complex as many players involved in an international market or a political treaty.

Examples of Players in Real Life

- In a corporate environment, the players might be various stakeholders, including employees, managers, and shareholders.
- In family decisions, players include family members whose preferences and choices impact one another.
- In international relations, players are countries with their governments making strategic decisions that affect global politics.

Strategies: Planning to Win

A strategy in game theory is defined as a complete plan of action a player will take in each possible state of the game. Depending on the game, a strategy can be simple (a single action) or complex (a series of actions or rules to follow). Types of Strategies:

- **Pure Strategies:** Involving a specific predetermined action regardless of the context.
- **Mixed Strategies:** Involving randomizing among different actions to potentially confuse opponents or to prevent them from being able to predict one's behavior.

- **Contingent Strategies:** These strategies depend on the history of the game thus far or on certain actions taken by other players.

Applying Strategies in Daily Life
- **Negotiating a Raise:** Your strategy might include preparing a list of accomplishments, timing your request after a successful project, or offering to take on new responsibilities in exchange for higher pay.
- **Deciding on a Route to Work:** You might choose a strategy based on traffic patterns, weather conditions, or the need to stop for errands along the way.

Payoffs: Understanding the Outcomes

In game theory, payoffs are the results players receive at the end of the game, based on the strategies they and others have chosen. Payoffs can include any type of benefit or loss, such as money, happiness, utility, or other forms of value. The nature of payoffs depends on the player's preferences and objectives. <u>Evaluating Different Types of Payoffs:</u>

- **Quantitative Payoffs:** These are measurable and often involve numerical values like profit, cost, or time.
- **Qualitative Payoffs:** These include outcomes that affect one's quality of life or satisfaction and can be subjective, like achieving a work-life balance or maintaining a happy relationship.
- **Immediate vs. Long-Term Payoffs:** Some strategies may provide immediate rewards, while others sacrifice short-term gains for long-term benefits.

Real-Life Examples of Payoffs
- **Investment Decisions:** The payoff might be the potential return on investment, considering the risk and the time value of money.
- **Choosing a Job:** Beyond salary, payoffs might include job satisfaction, career growth opportunities, and work-life balance.

Information Sets: The Role of Knowledge in Strategy

An information set in game theory represents what a player knows at any given point in the game, including the history of actions taken by all players up to that point. Information sets are crucial because they determine the strategies that a player considers viable. Categories of Information:

- **Perfect Information:** Players have complete knowledge of all previous actions and outcomes. Chess is a classic example where each player sees the entire board and all prior moves.
- **Imperfect Information:** Players have limited knowledge about some aspects of the game. For instance, in poker, players do not know the cards of their opponents.
- **Complete vs. Incomplete Information:** In games with complete information, all players know the strategies and payoffs available to each player. Incomplete information games involve unknown elements about the strategies or payoffs.

Applying Information Sets in Decision-Making
- **Buying a Car:** Your information set includes the cars available, their prices, reviews, and perhaps insider information from a friend in the car industry.
- **Career Strategy:** Information might consist of knowing the internal job postings, the company's growth strategy, and the decision-makers' preferences.

Integrating Game Theory in Everyday Life

Understanding the components of game theory—players, strategies, payoffs, and information sets—empowers you to approach daily challenges and opportunities with a strategic mindset. Whether you're making decisions about personal investments, negotiating at work, or even planning a vacation, considering these elements can lead to more favorable outcomes. Practical Tips for Using Game Theory:

1) Identify the Players: Recognize who is involved in each decision-making process and what their interests might be.

2) Determine Available Strategies: Think through your options and the potential responses from others.

3) Consider the Payoffs: Evaluate what you and others stand to gain or lose with each strategy.

4) Assess the Information: Consider how much you know and what you need to find out to make the best decision.

By systematically applying these concepts, you can enhance your ability to think critically and make decisions that not only benefit you but also take into account the broader dynamics at play. This approach doesn't just apply to high-stakes economic or business scenarios—it's equally relevant to the choices you face in everyday life.

Purpose and Use: How Game Theory is Applied in Various Fields

As you've begun to understand, game theory isn't just a theoretical construct used by economists and mathematicians—it's a versatile tool that can be applied in everyday life and professional fields alike. Whether you're negotiating a raise, strategizing over which route to take to work, or handling family dynamics, the principles of game theory offer valuable insights. Let's delve into how these principles are utilized across various domains, highlighting their significance and empowering you with the knowledge to apply them in numerous aspects of your life.

In Economics and Business

Game theory fundamentally transforms our understanding of market dynamics and competition. Consider a typical situation where two companies compete in the same market. Each company must decide on pricing, product launches, and marketing strategies without complete knowledge of the competitor's actions. Here, game theory provides a framework for predicting competitor behavior and formulating strategies that accommodate various potential actions by competitors.

For instance, the Nash Equilibrium, a fundamental concept in game theory, helps predict stable states in competitive environments where no participant can benefit by changing strategies while the other participants' strategies remain unchanged. This concept is crucial for you if you're in a role that requires strategic decision-making, such as marketing management or business development.

In Politics

In the political arena, game theory plays a critical role in election strategies, legislative negotiations, and international diplomacy. Political campaigns, for instance, can be seen as games where each candidate selects strategies that maximize their chances of winning, based on predictions of how their decisions will influence voter behavior.

Moreover, game theory explains the complex negotiations and alliances formed between political parties, especially in systems with multiple parties where no single party can achieve an absolute majority. Understanding these strategic interactions can enhance your ability to participate in or analyze political discussions and campaigns effectively.

In Social Sciences

Social scientists use game theory to study human behavior in various social contexts, including how individuals or groups cooperate or compete for resources. One interesting application is in the study of social norms and how they are enforced within communities. For instance, game theory can explain why certain behaviors are punished socially and others are rewarded, helping to maintain stability within the social group.

This understanding can be particularly beneficial if you find yourself in leadership or mediation roles where you need to foster cooperation and resolve conflicts within teams or groups.

In Biology

In evolutionary biology, game theory is used to explain how cooperation and competition affect the survival and reproduction of organisms. The Hawk-Dove game is a classic example used to explain animal conflict and territorial behavior. By understanding these strategies, biologists can predict which behaviors will become more prevalent over time based on their success.

If you're interested in environmental conservation or animal behavior, game theory provides a lens through which you can understand complex ecological interactions and the evolutionary significance of various behaviors.

In Technology and Computer Science

In the rapidly evolving field of technology, game theory is essential in areas such as network design, algorithm development, and cybersecurity. For example, in cybersecurity, game theory helps in designing strategies to handle threats and attacks by anticipating potential moves by hackers and implementing counter-strategies that minimize risk.

If you work in IT or cybersecurity, leveraging game theory can enhance your ability to develop more robust security measures and protect sensitive information effectively.

In Personal Decision-Making

On a more personal level, game theory has profound implications for everyday decision-making. From deciding whether to buy a car or a house to managing personal relationships, the strategic insights provided by game theory can help you navigate these decisions with more confidence and success. By evaluating the potential moves and payoffs of different choices, you can make more informed decisions that align with your long-term goals and values.

Types of Games: Introduction to cooperative vs. non-cooperative, static vs. dynamic, and complete vs. incomplete information games.

Game theory is a practical toolkit that you can use whether you're negotiating a raise, choosing the best route to work, or managing family dynamics. Together, let's delve into the distinctions between cooperative vs. non-cooperative games, static vs. dynamic games, and games with complete vs. incomplete information. By understanding these classifications, you'll gain deeper insights into the hidden strategies behind daily interactions and decisions, empowering you to lead a more thoughtful and strategic life.

Cooperative vs. Non-Cooperative Games

Games are typically categorized based on the nature of interaction between the players. The distinction between cooperative and non-cooperative games is particularly crucial as it defines how players align their strategies and whether they can form binding agreements.

Cooperative Games

In cooperative games, players can negotiate binding agreements that allow them to share strategies and rewards. These games focus on what coalitions will form, how the benefits of cooperation are distributed, and how group dynamics influence individual and collective decision-making. Imagine you're part of a project team at work where the success of the project depends on how well the team members collaborate, share resources, and align their efforts towards a common goal. The ability to negotiate, form alliances, and commit to group strategies can significantly impact the outcome.

In your personal life, cooperative game theory can explain scenarios like family financial planning or organizing a community event, where the success heavily relies on everyone working together towards a mutual benefit.

Non-Cooperative Games

Contrastingly, in non-cooperative games, binding agreements are not possible. Players must operate independently, making decisions based on individual strategies without the possibility to enforce cooperative agreements. Many competitive business environments, where companies vie against each other to maximize their own profits at the expense of others, typify non-cooperative games.

Consider a bidding war on a house you wish to buy. Each bidder independently decides how much to offer, often without knowledge of the others' bids. They must strategically decide their highest bid based on personal valuation, not on a cooperative agreement with other bidders.

Static vs. Dynamic Games

Another critical distinction in game theory is between static and dynamic games, which refer to the timing of players' moves.

Static Games

Static games are those in which all players make their decisions simultaneously, or at least without knowing the others' choices beforehand. These games often require you to predict others' decisions based on available information, without the ability to adjust your strategy based on others' actions as they occur.

An everyday example is rock-paper-scissors, where players simultaneously reveal their choices. In a more complex scenario, consider how you might simultaneously submit a job application among many candidates, trying to outshine others without knowing their qualifications or strategies.

Dynamic Games

Dynamic games allow players to make decisions at various points in time, often in response to the actions of others. This type of game is more reflective of real-life scenarios where decisions unfold in a sequence over time, and previous actions can influence future decisions.

For example, in a dynamic business negotiation, you might adjust your proposals based on the feedback or offers from the other party, engaging in a step-by-step strategy that evolves as the negotiation proceeds.

Complete vs. Incomplete Information Games

The amount of information available to players significantly affects their strategic choices in games. Games can be classified into those with complete information and those with incomplete information.

Complete Information Games

In games with complete information, every player knows the payoffs and strategies available to all other players. This transparency allows for well-informed decision-making. Classic board games like chess, where each player sees the entire board and knows all possible moves, typically fall into this category.

Incomplete Information Games

More commonly, especially in real-life scenarios, games involve incomplete information where players do not have full knowledge of other players' payoffs or strategies. Business markets often reflect this type, where companies must make strategic decisions based on limited information about competitors' actions, customer preferences, or market trends.

Real-World Applications and Insights

Understanding the types of games in game theory equips you with the knowledge to navigate complex personal, social, and professional scenarios. Whether it's deciding on a career move, negotiating a contract, or planning a family holiday, recognizing the type of game you are playing helps you choose the most effective strategy. It teaches you when to cooperate, when to compete, and how to adapt your strategies based on the dynamics of information flow and the nature of interactions with others.

By exploring these game types further, you'll be prepared to apply game theory principles more effectively, enhancing your ability to anticipate outcomes and make strategic decisions that lead to optimal results.

Personal Life Application: Making Everyday Choices, Conflict Resolution, Planning, and Decision-Making

As you navigate through life, have you ever considered how your daily choices could be influenced by game theory? This isn't just about big business decisions or high-stakes political maneuvers. It's about the everyday decisions that shape our lives, the conflicts we navigate, and the plans we make. Game theory isn't only for economists or strategists—it's a tool that you can use to enhance your personal decision-making. Let's dive deep into how the principles of game theory can illuminate the strategies behind your daily interactions, helping you make more informed choices that lead to better outcomes in your personal life.

Understanding Everyday Choices Through Game Theory

Every day, you face a myriad of decisions: What to eat for breakfast, which route to take to work, whether to spend or save your money, how to allocate your time after work, and so on. Each of these decisions might seem trivial, but they're all games that involve choices with various outcomes. By applying game theory, you can start to see these choices as part of a larger strategy for maximizing your overall happiness and efficiency.

Applying Game Theory to Routine Decisions

Consider your morning routine. You might think about whether to hit the snooze button or get up immediately. Game theory suggests evaluating the payoffs of each choice. If you get up immediately, your payoff is more time to enjoy breakfast and less rush, which could reduce stress. If you hit the snooze button, your immediate payoff is more rest, but the consequence might be a rushed morning or even being late, which could increase stress.

Game Theory in Financial Decisions

When it comes to spending or saving, game theory examines the strategies involved in achieving optimal financial well-being. For example, consider the decision to buy a coffee every morning versus saving that money. The payoff from saving could be financial security or a larger purchase down the line, while the payoff from buying coffee is immediate satisfaction. Game theory can help you balance short-term gratification with long-term goals by assessing the outcomes of different spending habits.

Conflict Resolution with Game Theory

Conflicts, whether with family, friends, or neighbors, can often be seen as non-cooperative games where the involved parties have opposing goals. However, by applying game theory, you can find strategies that lead to cooperative outcomes, turning potential conflicts into win-win situations.

Strategies for Family Disagreements

Imagine you and your partner disagree on vacation destinations. You want to go hiking in the mountains; they want to relax on a beach. Instead of arguing for your own preferences (a non-cooperative approach), game theory would suggest finding a compromise or an alternative that satisfies both parties' preferences to some extent. Perhaps a coastal town with access to both beaches and hiking trails could be an agreement that maximizes both of your happiness.

Resolving Conflicts Among Friends

Game theory also applies to planning events with friends, where conflicting schedules and interests can complicate decisions. Instead of letting one person decide, you can use game theory to propose multiple options and let everyone rate their preferences. This approach, often used in voting theory, can help achieve a consensus that considers everyone's top choices, minimizing dissatisfaction.

Planning and Decision-Making with Game Theory

Planning involves projecting future needs and deciding on the best courses of action to meet those needs. Game theory enhances planning by considering the interactions between different decisions over time.

Using Game Theory for Personal Goals

Setting personal goals, such as achieving fitness, learning new skills, or career advancement, involves planning and often, strategic interactions with others. For instance, if you're planning to get fit, game theory suggests considering how your schedule interacts with other commitments. A strategy might involve negotiating time slots with family members so you can work out while they are engaged in their activities, ensuring no conflicts and maintaining harmony.

Long-Term Decisions: Retirement Planning

Game theory can profoundly impact long-term decision-making, such as retirement planning. The decisions you make today, from saving to investing, will affect your future payoffs. Game theory teaches you to evaluate various strategies based on expected outcomes and risks, helping you choose a plan that best secures your future financial stability.

Career Application: Job Selection, Networking Strategies, Career Advancement

Whether you're at the crossroads of choosing a new job, strategizing over networking, or contemplating the next steps in your career advancement, game theory offers invaluable tools to help navigate these decisions. Let's delve into how the principles of game theory can clarify and enhance the strategic choices you make in your career, guiding you toward more informed and effective outcomes.

Job Selection: Choosing the Right Path

Choosing a job is one of the most significant decisions you'll make in your life. It's not just about the immediate benefits such as salary and perks; it involves long-term implications for your career path, personal development, and satisfaction.

Evaluating Offers Through Game Theory

Imagine you have multiple job offers. Each offer has different benefits and drawbacks concerning salary, job role, career growth potential, company culture, and location. Game theory encourages you not just to evaluate each offer in isolation but to consider how each decision might play out in various future scenarios. What are the payoffs from each potential choice? Which option gives you the best balance of immediate satisfaction and long-term growth?

Using a decision matrix, you can assign values to the benefits and drawbacks of each offer based on your personal priorities and goals. This process helps clarify which job aligns best with your long-term career aspirations, optimizing your professional trajectory.

Strategic Moves in Negotiations

Once you've selected your preferred job, game theory also offers strategies for negotiating your contract. Understand the negotiation as a game where both you and your potential employer have strategies and goals. Your employer wants to secure your talent at a reasonable cost, while you want to maximize your compensation and benefits. Recognizing the employer's constraints and needs can guide you to propose a deal that appeals to both parties, thus increasing the chances of a successful negotiation.

Networking Strategies: Building Valuable Connections

Networking is not just about collecting contacts; it's about building valuable relationships that can open doors to opportunities and support your career development. Game theory provides a framework for understanding the dynamics of networking events and interactions.

Creating Win-Win Situations

In any networking scenario, think of each interaction as a mini-game where the objective is to create value for both parties. Approach networking with the mindset of mutual benefit: what can you offer, and what can you gain from each interaction? This perspective helps you build more meaningful and sustainable relationships.

For instance, if you meet someone at a conference who works in a role or company you're interested in, consider how you can make the interaction beneficial for them as well. Can you share insights from your experience, offer assistance with a project, or connect them with another valuable contact? By doing so, you're more likely to forge a connection that reciprocates in value, perhaps by way of advice, mentorship, or future job opportunities.

Career Advancement: Navigating the Ladder

Advancing in your career is a strategic challenge that involves identifying opportunities, enhancing your skills, and sometimes, making bold moves.

Analyzing Your Career Path

Use game theory to analyze your current position and potential career paths just as you would analyze moves in a game. What are the potential outcomes of pursuing a promotion, switching departments, or even changing companies? Each career move can be seen as a strategic decision with associated risks and rewards.

Evaluate the landscape of your industry: Are there emerging sectors or roles where you could quickly move up? Are there moves you could make that might block or enhance your competitors' (colleagues') strategies for advancement?

Timing and Strategic Moves

In career advancement, timing can be everything. Game theory helps you decide the optimal time to make a move. Consider the state of your company, industry trends, and your personal circumstances. Sometimes, advancing your career might mean taking a risk at an opportune moment, such as leading a high-stakes project that could significantly elevate your visibility within the company.

Business Application: Competitive strategies, pricing strategies, partnership decisions

Competitive Strategies: Outmaneuvering the Competition

In the competitive arena of business, every company seeks to establish a stronghold in its market while keeping competitors at bay. Game theory provides a robust framework for analyzing competitive interactions, predicting competitor moves, and strategizing accordingly.

Anticipating Competitor Moves

Imagine you're the head of a company that's about to launch a new product. Before you make your move, it's crucial to anticipate how your competitors might react. Will they lower their prices, improve their own products, or perhaps launch an aggressive marketing campaign? Game theory encourages you to think several steps ahead, just like in chess, allowing you to prepare counter-strategies that protect your market position and potentially exploit competitors' weaknesses.

Creating Barriers to Entry

Using game theory, you can also strategize on creating barriers to entry in your market, making it difficult for new competitors to emerge. This might involve strategies like scaling up production to lower costs, securing favorable supplier contracts, or innovating rapidly to stay ahead of the technological curve. Each of these moves requires careful consideration of how competitors and potential entrants will respond, ensuring that your strategies are not only robust but also adaptable.

Pricing Strategies: Maximizing Profitability

Pricing is an area where game theory is particularly powerful, helping you determine the optimal price points for your products or services based on market dynamics and consumer behavior.

Price Wars and Their Implications

Consider a scenario where you're competing on price with several other firms in a highly competitive market. Engaging in a price war might seem like a good short-term strategy to increase market share, but game theory analysis could show that this will hurt profitability in the long term. Instead, it might be more advantageous to compete on quality or service, differentiating your offerings from competitors, thus allowing for higher price points without sacrificing sales.

Dynamic Pricing Models

Game theory also supports the implementation of dynamic pricing models where prices adjust based on real-time supply and demand. Airlines and hotels often use these models, but they can be applied in various industries. Analyzing consumer response patterns and competitor pricing through game theory allows businesses to optimize their pricing strategies dynamically, maximizing revenue potential.

Partnership Decisions: Choosing the Right Allies

In business, forming the right partnerships can be as critical as choosing the right business strategies. Game theory helps you evaluate potential partnerships, foreseeing the benefits and risks, and guiding negotiation tactics.

Evaluating Partnership Opportunities

When considering a partnership, game theory prompts you to analyze not only what each party brings to the table but also their potential future actions. Will this partnership give your partner too much power over your operations? Could they become a competitor? What contingencies can you build into the partnership agreement to protect your interests? These are crucial questions that game theory helps you answer.

Strategic Alliances

In forming strategic alliances, game theory encourages a cooperative approach where all parties are incentivized to work together towards common goals. It helps outline how alliances can be structured to ensure mutual benefit, maintain balance, and avoid conflicts. For instance, a co-marketing agreement between two firms should be designed so that both firms benefit roughly equally from the arrangement, preventing any imbalance that could strain the relationship.

Sports Application: Game Planning, In-Game Decisions, Player Negotiations

Game Planning: Strategizing for Victory

Game planning in sports involves preparing strategies before the game even starts. Coaches and players analyze their opponents' strengths and weaknesses, past performances, and likely tactics to develop a game plan that maximizes their chances of winning.

Pre-Game Analysis and Strategy Formulation

Imagine you're a soccer coach preparing for an important match. Game theory can help you decide whether to adopt an aggressive attacking strategy or a more conservative defense-oriented game plan. By modeling the game, you can predict possible outcomes based on different strategies, taking into account not only your team's capabilities but also the opponent's known behaviors and responses.

For example, if the opposing team is known for their slow starts but strong finishes, you might plan an aggressive start to capitalize on their initial sluggishness, aiming to secure an early lead that puts psychological pressure on them for the rest of the match.

Optimizing Team Formation and Player Roles

Game theory also aids in deciding the best team formation and player roles based on the strategic goals. Different formations can be modeled as different strategies in a game, with each having particular strengths and weaknesses against various opposing strategies. This analytical approach allows you to tailor your team's strengths to best exploit the weaknesses of the opponents, adapting your strategy to each new game situation.

In-Game Decisions: Dynamic Strategies and Tactics

During the game, coaches and players must make constant decisions that can affect the outcome. Game theory provides a framework for making these decisions more strategically, adapting to the unfolding dynamics of the game.

Real-Time Decision Making

Consider a basketball coach deciding when to call a timeout or substitute players. Game theory can inform these decisions by considering them as moves in a game, where each decision can change the momentum or counter an opponent's strategy effectively. For example, introducing a fresh player who has a good track record against a particular opponent can be a game-changing move, altering the dynamics and forcing the opponent to adjust their strategy.

Adapting to Opponent Moves

In-game decisions also involve adapting to the strategies and tactics employed by the opponents. Game theory helps in anticipating possible moves by the opponents and planning counter-moves in advance. This kind of strategic foresight is critical in sports like tennis or chess, where anticipating the opponent's next move and countering effectively can lead to dominance in the game.

Player Negotiations: Contract Talks and Transfers

Player negotiations are a critical aspect of sports management, involving discussions about contracts, salaries, and transfers. These negotiations can be complex, involving multiple parties with different goals and expectations.

Modeling Negotiations as a Game

In player negotiations, game theory models the negotiation process as a bargaining game where each party has different strategies and outcomes they are willing to accept. For instance, a football player's agent might negotiate with several clubs, each offering different terms. Game theory helps in evaluating these offers, considering not only the financial aspects but also the player's career growth, the competitiveness of the league, and other personal preferences.

Strategic Use of Information

Information plays a crucial role in negotiations. Game theory emphasizes the strategic use of information — revealing or withholding certain information can influence the negotiation process. For example, an agent might leverage interest from multiple teams as a strategy to secure a better deal, while teams might use information about a player's health or recent performance to negotiate down the terms.

Social Application: Group Dynamics, Event Planning, Social Bargaining

Understanding Group Dynamics Through Game Theory

Group dynamics can be complex, with various individuals bringing different perspectives, interests, and goals into social interactions. Game theory helps to decode these dynamics by modeling groups as players in a game, where each member's decisions impact the outcomes for the entire group.

Analyzing Social Interactions as Strategic Games

Imagine you're part of a local club or community group facing a decision on a new initiative. Each member has their own preferences and ideas about what should be done. Game theory enables you to analyze these situations by considering each member's potential choices and the consequences of these choices on group harmony and the success of the initiative. It guides you to look for equilibrium solutions where the group can converge on a decision that, while not perfect for everyone, is acceptable to all.

Fostering Cooperation and Resolving Conflicts

In any group, conflicts might arise due to differing interests or misunderstandings. Game theory can provide strategies for conflict resolution by identifying mutually beneficial outcomes and suggesting negotiation tactics that promote compromise. For example, if two members of a community group disagree on the budget allocation for an event, game theory could help outline a compromise that allocates funds in a way that meets the most crucial needs of both parties, thereby maintaining group cohesion.

Event Planning: Maximizing Satisfaction and Participation

Planning social events, whether large public gatherings or small private parties, often involves complex decision-making where the preferences of multiple stakeholders must be considered. Game theory helps you navigate these challenges by treating event planning as a strategic game where different outcomes have different payoffs.

Optimizing Event Outcomes

Consider the planning of a community festival where multiple elements like venue, entertainment, food, and logistics need to be coordinated. Game theory allows you to model different scenarios—such as venue choice affecting attendance numbers—and helps you make decisions that maximize overall participant satisfaction. By analyzing how different choices impact attendee enjoyment and participation, you can strategically select options that enhance the overall success of the event.

Balancing Diverse Preferences

In situations where community members have diverse and sometimes conflicting preferences, game theory provides a framework for finding a balance that respects these differences while still achieving the goals of the event. For instance, when selecting entertainment options, game theory can help you evaluate which choices will please the greatest number of attendees while still staying within budget constraints.

Social Bargaining: Negotiating Win-Win Outcomes

Social bargaining involves negotiations that occur in everyday interactions, such as dividing responsibilities within a household, setting rules within a community, or even determining social norms. Game theory excels in these areas by providing strategies that ensure negotiations are fair and beneficial to all parties involved.

Enhancing Negotiation Tactics

Imagine you're negotiating with your neighbors to establish a shared space for community use. Each neighbor has different ideas and preferences about how the space should be used. Game theory helps you approach these negotiations strategically, ensuring that you consider not only your own preferences but also the potential counteroffers and priorities of your neighbors. It guides you in crafting proposals that anticipate and incorporate the reactions of others, leading to agreements that are more likely to be accepted and sustained.

Creating Equitable Solutions

In family settings, such as deciding on how to divide household chores among family members, game theory can help design fair division rules that consider the abilities, preferences, and time constraints of each family member. By treating this as a cooperative game, you can develop solutions that everyone feels are equitable, thus minimizing conflicts and enhancing family harmony.

Environmental Application: Resource Management, Conservation Efforts, Community Action

Resource Management: Allocating Resources Strategically

Effective resource management is crucial in ensuring the sustainable use of our planet's limited resources. Game theory provides a framework for analyzing situations where multiple parties compete for the same resources, helping to devise strategies that maximize the overall utility while promoting conservation.

Managing Water Resources

Consider the management of water resources in a region where multiple stakeholders, including farmers, municipalities, and industrial users, compete for water access. Game theory helps you model this situation as a cooperative game, where each stakeholder's payoff depends on not only their own water usage but also on how much water is left for others. Strategies derived from game theory can encourage stakeholders to share water resources in a way that supports both individual needs and overall regional sustainability.

Forest Conservation Efforts

In forest conservation, game theory can address the conflict between logging interests and conservation goals. By modeling the forest as a shared resource, game theory helps stakeholders understand the long-term benefits of sustainable logging practices versus the short-term gains from deforestation. This approach can lead to agreements that allow for sustainable use of the forest while preserving its ecological integrity for future generations.

Conservation Efforts: Promoting Biodiversity and Sustainability

Conservation efforts often involve complex decisions about how to best preserve biodiversity while meeting human needs. Game theory provides insights into how different conservation strategies can lead to varying outcomes based on the behavior of involved parties.

Protecting Endangered Species

Imagine you are part of a global initiative to protect an endangered species. Game theory can help in designing strategies that involve multiple countries and stakeholders. By understanding the incentives for different parties to participate in or oppose conservation efforts, game theory can guide the creation of incentive structures that encourage cooperation and effective conservation actions.

Community-Led Conservation Projects

In community-led conservation projects, such as the establishment of a marine protected area, game theory assists in negotiating agreements among local stakeholders, such as fishermen, tourism operators, and conservation groups. By identifying potential conflicts and cooperative strategies, game theory helps communities reach consensus on how to manage and protect their natural resources effectively.

Community Action: Mobilizing for Environmental Change

Community action is vital for driving environmental change at the local and global levels. Game theory offers tools to mobilize communities, foster collaboration, and overcome public goods dilemmas where individual incentives might not align with collective environmental goals.

Organizing Clean-Up Efforts

Consider organizing a community clean-up event. Game theory can help you understand the factors that motivate volunteers to participate and how to structure rewards and recognition to maximize participation. Strategies can be designed to appeal to different community segments, ensuring broad involvement and making the event a success.

Advocating for Policy Changes

When advocating for environmental policy changes, game theory provides a strategic framework for understanding the positions and potential responses of different stakeholders, including politicians, industries, and the public. This understanding can shape more effective lobbying strategies that consider not just direct arguments for change but also the strategic positioning and counter-moves of opponents.

Reflection Questions

1) Everyday Decisions: Reflect on a recent decision you made, either at work or in your personal life. How could viewing this decision through the lens of game theory (considering players, strategies, payoffs, and information sets) alter your understanding of the outcome or your approach to similar decisions in the future?

2) Strategic Interactions: Think of a scenario in your career or personal life where the outcome depended not just on your actions but also on the actions of others. What type of game (cooperative vs. non-cooperative, static vs. dynamic) does this scenario best resemble? How might your strategy change if you could redesign the game rules?

3) Applications of Game Theory: Choose one of the applications of game theory listed (e.g., environmental resource management, sports game planning, business competitive strategies). Can you think of an example from recent news or your own experience where principles of game theory were evident? How did the understanding of game theory components potentially influence the outcomes?

Section 3: Understanding Rational Decision Making

In this section we are going to dive deep into the concept of rationality, a cornerstone of decision-making in game theory and beyond. Rational decision-making is an essential skill that influences outcomes in every aspect of life, from personal finances and health choices to strategic business negotiations and policy-making. Through this exploration, we will uncover what it means to make a rational decision, examine the criteria that define rationality, and discuss the Expected Utility Theory, which provides a foundational framework for understanding how choices are made.

We will also address the limitations and challenges that arise in rational decision-making—recognizing that real-world scenarios often present complexities that can deviate from theoretical models. By understanding these limitations, you can better navigate the practical challenges encountered in various decision-making environments.

Furthermore, this chapter applies the principles of rational decision-making across a spectrum of real-world applications:

- In **personal life**, such as financial planning and health choices, where making rational decisions can significantly impact long-term wellbeing.
- In **career development**, where strategic planning and skill development play crucial roles in navigating professional landscapes.
- In **business scenarios**, where rational strategies in negotiations, risk management, and long-term planning are vital for success.
- Within **government frameworks**, where policy-making and public administration require a balance of rationality and adaptability to manage societal needs.
- In the **legal field**, where litigation strategies and negotiations must be grounded in rational analysis to effectively represent and protect client interests.
- And in **environmental management**, where rational decision-making supports sustainable policy design and resource allocation.

We will guide you through understanding how rational decision-making is applied in these fields, providing you with the tools to make informed, effective decisions. By the end of it, you will have a comprehensive understanding of the dynamics of rational decision-making and its critical role in shaping successful outcomes across various domains of life. Let's begin our journey into the rational decision-making process, enhancing your ability to apply these concepts thoughtfully and effectively in both everyday situations and complex professional contexts.

Defining Rationality: What It Means to Make a "Rational" Decision

As we navigate through life's myriad choices, understanding the essence of rational decision-making can greatly enhance our ability to make choices that not only serve our immediate needs but also align with our long-term goals. Whether you're deciding on a career move, planning your finances, or simply choosing where to dine, the concept of rationality is key to navigating these decisions effectively and efficiently.

The Essence of Rational Decision-Making

Rationality involves making choices that are consistent with one's own goals and preferences, using a logical and systematic approach. It's about aligning your decisions with clear, well-defined objectives and having a coherent strategy to achieve these objectives.

Core Principles of Rationality

- **Consistency**: Your decisions should be consistent over time and not contradict each other. For example, if you decide saving money is important today, you should not impulsively spend tomorrow.
- **Maximization**: At the heart of rationality is the goal to maximize your **utility**, which means <u>making choices that provide the greatest benefit or satisfaction based on your preferences and values.</u>
- **Systematic Analysis**: Rational decision-making involves a systematic evaluation of the possible options, considering all available information and potential outcomes before making a choice.

Understanding Preferences and Goals

The foundation of rationality is knowing what you want—understanding your preferences and setting clear, achievable goals. Your preferences are the desires or needs that drive your decision-making. These can range from basic needs like food and shelter to more complex desires such as career fulfillment or personal growth. Rational decision-making requires you to be clear about these preferences and to order them in terms of priority. This clarity helps in making decisions that truly reflect what is most important to you. Goals give direction to your rational decisions. They transform vague aspirations into concrete targets. For instance, instead of simply wishing to be wealthier, a rational approach involves setting a specific goal, such as increasing your savings by a certain amount over a year. This specificity makes it possible to plan effectively and measure your progress.

Applying Logic and Evidence

Rational decision-making relies heavily on the use of logic and empirical evidence. This means basing your decisions on facts, data, and logical reasoning rather than emotions or intuition.

Logical reasoning involves evaluating relationships among different facts and variables, deducing conclusions from premises, and ensuring that your decisions follow logically from the information available. For example, if you know that investing in education can increase job opportunities (a premise), you might conclude that pursuing further education is a rational choice if your goal is career advancement.

Using empirical evidence means basing your decisions on data, observations, and experiential learning. For instance, if data shows that a certain investment has consistently yielded high returns, a rational decision might be to invest in that asset class, assuming it aligns with your risk preference and financial goals.

Challenges in Rational Decision-Making

While the principles of rationality are straightforward, applying them can be challenging. Cognitive biases, emotional influences, and imperfect information often skew our ability to make truly rational choices.

Cognitive Biases

These are consistent ways in which judgment deviates from what is normal or rational. Biases like overconfidence, anchoring on specific information, or being swayed by the status quo can lead you to make decisions that aren't truly aligned with your rational self-interest.

Emotional Influences

Emotions play a critical role in decision-making but can sometimes lead to choices that defy rational analysis. Fear, excitement, or stress can cloud judgment and lead to decisions that might not be in your best long-term interest.

Imperfect Information

Often, we have to make decisions without having all the necessary information. Uncertainties about the future, incomplete data about alternatives, or unknown factors can all complicate the decision-making process.

Criteria for Rationality: How Rationality is Evaluated in Different Scenarios

As we continue our exploration into rational decision-making, it becomes essential to understand how rationality is evaluated. This involves dissecting the criteria that determine whether a decision is rational. These criteria not only guide us in assessing our own choices but also help in understanding the decisions made by others around us, whether in personal life, at work, or in broader societal contexts. Let's delve into the criteria for rationality, exploring how these standards can be applied to various scenarios to ensure decisions are both effective and aligned with overarching goals.

Understanding the Criteria for Rational Decision-Making

Rational decision-making is not just about making choices; it's about making the right choices based on a logical and structured evaluation of available information and potential outcomes. To better navigate this complex landscape, let's consider the fundamental criteria that guide rationality.

1. Consistency with Goals and Preferences

The first criterion for rationality is that decisions must consistently align with the predefined goals and preferences. This alignment ensures that every choice you make takes you a step closer to your desired outcomes.

Scenario: Career Advancement

Suppose you're deciding whether to pursue a new job opportunity. The rational choice would typically be the one that aligns with your long-term career goals, such as advancing your professional skills or moving into a leadership role. Evaluating how well each job aligns with these goals can help determine the most rational decision.

2. Optimality of Outcome

Rational decisions should optimally balance risk and reward, aiming for the outcome that provides the maximum benefit with acceptable risk levels.

Scenario: Investment Decisions

When choosing an investment, the rational decision is often the one that offers the best possible return for the lowest acceptable risk, according to your risk tolerance and financial goals. Utilizing tools like risk-return models can aid in making a decision that best fits these criteria.

3. Efficiency in Resource Utilization

Efficiency is a key criterion in rational decision-making, ensuring that resources (time, money, effort) are used in the most effective way to achieve desired outcomes.

Scenario: Project Management

In managing a project, rational decisions involve allocating resources in a manner that maximizes productivity without overuse or waste. This might include tasks like delegating responsibilities according to team members' skills or scheduling work in phases to ensure optimal progress.

4. Based on Reliable Information and Sound Logic

Rational decisions should be grounded in reliable information and logical reasoning, ensuring that choices are informed and reflective of the reality of the situation.

Scenario: Health Choices

Choosing a treatment option for a medical condition should be based on sound medical information and advice. Rationality involves weighing the effectiveness of different treatments based on clinical evidence and personal health circumstances.

Applying Rationality Criteria in Diverse Situations

Understanding and applying these criteria can vary significantly across different life scenarios. Each situation demands a tailored approach to evaluate rationality effectively.

Personal Life: Financial Planning

In personal financial planning, rationality means making budgeting and investment decisions that align with your financial security and lifestyle goals. It involves regular review and adjustment of your financial plan to ensure it remains optimal as your personal circumstances and market conditions change.

Business Environment: Strategic Planning

In a business context, rational decisions are those that contribute to the company's strategic objectives, such as growth, profitability, and market position. This involves not just choosing the right strategies but also continually assessing the business environment and adjusting plans as needed to maintain strategic alignment.

Social Context: Community Action

In community or social settings, rational decisions often revolve around achieving the greatest good for the community. This might involve organizing community resources for collective benefits, such as developing local parks or community health initiatives, ensuring that actions taken are efficient and beneficial for the majority.

Expected Utility Theory: Basic Principles Behind Rational Choice Theory

Expected Utility Theory is a pivotal concept in rational choice theory that serves as a fundamental tool in both economic decision-making and everyday life choices. As we delve into this topic, you'll discover how this theory helps to frame rational decisions in terms of maximizing expected utility—a measure of the satisfaction or value derived from specific outcomes. Whether you're deciding on investment options, choosing between job offers, or even planning your next vacation, understanding the principles of Expected Utility Theory can significantly enhance your decision-making processes.

What Is Expected Utility Theory?

Expected Utility Theory is a mathematical framework used to model decision-making under uncertainty. It posits that individuals choose between risky or uncertain prospects by comparing their expected utility values, which represent the sum of the utilities of all possible outcomes, weighted by the probabilities of these outcomes.

Fundamental Assumptions

Completeness: Every possible choice can be ranked; you have clear preferences that allow you to say whether you prefer one outcome over another, or are indifferent between them.

Transitivity: Your preferences are consistent across choices; if you prefer option A over option B, and option B over option C, then you must prefer option A over option C.

Independence: Your preference between options should remain consistent, even when they are part of a broader combination of prospects.

Continuity: Small changes in the probability of a particular outcome should not lead to abrupt changes in your preferences between different options.

Applying Expected Utility Theory in Everyday Decision-Making

To understand how Expected Utility Theory applies to everyday decisions, let's consider a simple example. Imagine you are faced with a choice between receiving a guaranteed $50 or having a 50% chance to win $100 and a 50% chance of winning nothing.

Calculating Expected Utility

If you value monetary gains linearly (meaning the utility is directly proportional to the amount), the expected utility of the gamble is 0.5 * $100 + 0.5 * $0 = $50. The expected utility of the guaranteed $50 is simply $50. If these values are the same, and you are risk-neutral, you may be indifferent between the gamble and the guarantee. However, if you are risk-averse, the guaranteed $50 might provide higher utility due to the absence of risk.

The Role of Risk Preferences in Expected Utility

Risk preferences play a crucial role in Expected Utility Theory. Individuals can be:

Risk-Averse: Preferring a certain outcome over a gamble with an equal expected outcome. For instance, preferring a guaranteed job offer over the chance of a potentially better position that is not guaranteed.

Risk-Neutral: Indifferent between sure things and gambles of equal expected value.

Risk-Seeking: Preferring gambles over sure things even when the expected values are equal, possibly valuing the thrill or other aspects of taking risks.

Decision-Making in Complex Scenarios

Expected Utility Theory not only applies to simple gambles but also helps in complex decision-making scenarios involving multiple options and outcomes.

Investment Decisions: When deciding how to invest your savings, Expected Utility Theory helps you evaluate various investment options by considering the potential returns (benefits) and the risks (probabilities of different returns) associated with each option. For example, stocks have higher potential returns but greater risk compared to bonds, which are generally safer but offer lower returns. Your choice will depend on your risk preference and the expected utility of each investment option.

Career Choices: Choosing a career path can also be analyzed using Expected Utility Theory by weighing the potential benefits (salary, job satisfaction) against the risks (job market volatility, educational requirements). Calculating expected utilities can help clarify which career might yield the greatest overall satisfaction.

Healthcare Decisions: When facing medical decisions, especially under uncertainty regarding outcomes, Expected Utility Theory can assist in choosing treatments by weighing the potential health benefits against the risks and side effects.

Limitations of Expected Utility Theory

While powerful, Expected Utility Theory is not without its limitations. It assumes that individuals have the ability to process all relevant probabilities and outcomes, which is not always practical in real-world situations. Furthermore, it cannot always account for the intrinsic values or emotional factors that people may consider in their choices.

Limitations and Challenges: Recognizing Where Rational Decision-Making Can Fall Short

While the principles of rationality provide a robust framework for making choices that maximize utility and align with our goals, there are several scenarios where this approach can fall short. Understanding these limitations is crucial for you to navigate complex decision-making landscapes more effectively, allowing you to anticipate potential pitfalls and adapt your strategies accordingly. Let's delve into these challenges and explore how awareness of them can enhance your decision-making process in both personal and professional contexts.

Cognitive Limitations and Biases

One of the primary challenges in rational decision-making arises from our own cognitive limitations and inherent biases, which can skew our perception and analysis of information.

Cognitive Overload

Decision-making often requires the processing of a large amount of information. However, our cognitive resources are limited. When faced with complex decisions involving numerous variables and outcomes, you may experience cognitive overload, which can hinder your ability to assess each option rationally. This overload can lead to shortcuts in thinking, known as heuristics, which do not always lead to optimal decisions.

Inherent Biases

Biases in decision-making are systematic deviations from logic or rationality in judgment. These can include:

- **Confirmation Bias**: The tendency to favor information that confirms your existing beliefs, leading to skewed decision-making. For example, when deciding on a new market strategy, you might give undue weight to data that supports your preferred approach while disregarding contrary evidence.
- **Loss Aversion**: A phenomenon where the pain of losing is psychologically about twice as powerful as the pleasure of gaining. In financial decisions, this can make you irrationally risk-averse, potentially missing out on opportunities with high rewards because the fear of loss is overwhelming.
- **Anchoring Effect**: The tendency to rely too heavily on the first piece of information you hear when making decisions. For instance, initial price points can influence your perception of subsequent offers in negotiations, regardless of their actual value.

Emotional Influences

Rational decision-making assumes that choices are made based on logical assessment of outcomes. However, emotions play a significant role in many decisions.

Emotional Decision-Making

Decisions can be heavily influenced by emotions, which can sometimes lead to choices that defy rational analysis. For instance, a decision about a job change might be swayed more by your feelings about your current workplace environment than by an objective evaluation of career benefits. Emotions such as fear, excitement, or stress can dramatically affect decisions, from financial investments to personal relationships.

Impact on Risk Perception

Emotions can also alter your perception of risk and reward. In high-pressure situations, such as during a crisis or in high-stakes negotiations, heightened emotions can lead to riskier decisions that might not be made under normal circumstances.

Information Asymmetry

Rational decision-making requires access to relevant and accurate information. However, information asymmetry can pose a significant challenge.

Incomplete Information

Often, you have to make decisions without access to all the necessary information. This incomplete information can lead to suboptimal choices. For instance, in a real estate investment, hidden factors such as future area developments or zoning laws can impact the decision-making process, potentially leading to unforeseen complications.

Misinformation

In the era of information overload, misinformation can easily find its way into your decision-making process. The ability to critically evaluate the accuracy and relevance of information is crucial, yet distinguishing between reliable and unreliable information can be challenging.

Practical Constraints

Finally, practical constraints often limit the applicability of rational decision-making.

Time Constraints

Decisions often need to be made quickly, especially in fast-paced environments like financial markets or emergency situations. These time constraints can prevent a thorough rational analysis, forcing you to rely on intuition or incomplete information.

Resource Limitations

Resources, whether time, money, or expertise, are often limited. These limitations can constrain your options and force decisions that are less than optimal but are the best under the circumstances.

Examples of Rational and Non-Rational Decision-Making in Game Theory

Part 1: Rational Decision-Making According to Game Theory

Rational decision-making in game theory is characterized by choices that maximize an individual's utility based on their preferences and the available information. Here are five examples crafted into narratives:

1. The Stock Market Investment Decision Jake is an investor assessing his options in the stock market. He uses mixed strategies to diversify his portfolio, choosing stocks that balance risk and return optimally according to personal risk tolerance and market data. By considering historical performance, current market trends, and economic indicators, Jake makes decisions that maximize his expected utility, embodying the rational actor model in game theory.

2. Corporate Strategy in Oligopolistic Markets A group of companies operates in an oligopolistic market where pricing decisions of one affect the others. Firm A, led by CEO Ms. Thompson, decides to use game theory to determine a pricing strategy. By analyzing competitors' responses and potential market changes, Firm A sets a price that maximizes profits without inciting a price war, achieving a Nash Equilibrium where no player can benefit by changing strategies unilaterally.

3. Negotiation of Salary in Job Offers Sarah receives a job offer and enters salary negotiations. Understanding the principles of Nash Bargaining, she evaluates her value, the compensation framework of the company, and market rates for similar positions. Sarah and her potential employer reach an agreement that maximizes both parties' payoffs, reflecting a rational decision-making process where both sides adjust their strategies to achieve an optimal outcome.

4. Resource Allocation in Public Goods The local government is deciding how to allocate its budget across public services such as parks, roads, and schools. By employing mechanism design, officials create a plan that maximizes social welfare, considering the strategic interactions and preferences of different community groups. The decision-making process strategically allocates resources to benefit the entire community, showcasing rationality in public administration.

5. Strategic Voting in Elections During an election with multiple candidates, voter Emily faces a strategic decision. Instead of voting for her preferred but less popular candidate, she votes for a more viable candidate who can realistically win against a less favored opponent. Emily's choice illustrates strategic voting—a rational decision to maximize the utility of her vote, ensuring the best possible outcome under the circumstances.

Part 2: Non-Rational Decision-Making in Game Theory

Non-rational decision-making occurs when choices do not maximize expected utility. These decisions are often influenced by irrational biases, lack of information, or emotional responses, as illustrated in the following narratives:

1. Panic Selling in Stock Markets During a sudden market downturn, investor Tom reacts impulsively by selling his stocks to avoid further losses, driven by fear and herd behavior. This decision, made in a state of panic, leads to selling at low prices and realizing losses, which contradicts the rational decision-making models suggested by game theory that advocate for holding investments long-term for potential recovery.

2. Predatory Pricing to Eliminate Competitors Firm B, dominated by a confident CEO, engages in predatory pricing to drive competitors out of the market, setting prices unsustainably low. While initially seeming like a strategic move, this leads to significant long-term losses due to legal consequences and a weakened market structure once competitors are gone. This decision shows non-rational behavior influenced by overconfidence and a focus on short-term gains over long-term profitability.

3. Overbidding in Auctions At an art auction, collector Ann becomes emotionally attached to a painting. Caught in the competitive spirit of the auction, she continues to bid beyond the painting's objective market value. Her decision to win at any cost, driven by emotional investment rather than rational evaluation, exemplifies non-rational decision-making.

4. Under-investment in Public Goods In a small town, residents decide not to contribute to the maintenance of local parks, expecting others to cover the necessary costs. This leads to deteriorated park facilities, ultimately reducing the quality of life for all, including the non-contributors. This behavior, known as the free-rider problem, illustrates non-rational decision-making where individuals avoid personal costs at the expense of collective benefit.

5. Refusing to Negotiate in Hostage Situations In a critical hostage situation, government officials decide not to negotiate with the kidnappers to avoid setting a precedent. This decision is made to deter future kidnappings but at the cost of potentially saving lives immediately. The choice, driven by a policy stance rather than the immediate maximization of welfare, shows how emotional or ethical considerations can lead to non-rational decisions in high-stakes environments.

Personal Life Application: Financial Planning, Health Choices, Risk Assessment

Here, we will delve into how rationality informs decisions that impact your financial stability, health, and personal risk management. Understanding how to apply rational decision-making processes to these fundamental areas can greatly enhance your life quality by aligning your choices more closely with your long-term goals and values.

Financial Planning: Navigating Economic Choices Rationally

Financial planning is perhaps where rational decision-making can be most impactful. Here, you are often faced with decisions that not only have immediate consequences but also affect your future financial stability.

Creating a Budget

The first step in rational financial planning is often creating a budget. This involves a detailed assessment of your income versus your expenditures. A rational approach to budgeting requires not just tracking where your money goes but planning and optimizing how you use every dollar to achieve both short-term satisfaction and long-term financial goals.

Strategy: Start by listing all your sources of income and all your necessary expenses, such as rent, utilities, and groceries. Then allocate funds to savings and investments before budgeting for discretionary spending. This method ensures that you prioritize financial security and goal attainment.

Investment Choices

Investing is another critical area in financial planning where rational decision-making is essential. Choosing where and how much to invest should be based on a thorough risk assessment and an understanding of potential returns, aligned with your financial goals and risk tolerance.

Application: Utilize tools like diversification to minimize risk and expected utility theory to balance potential gains against risks. For instance, if you're risk-averse, you might lean towards bonds and mutual funds, rather than more volatile stocks, aligning your portfolio with your comfort level and long-term objectives.

Health Choices: Rational Decisions for Well-being

Your health decisions are often complex and laden with emotional and sometimes contradictory information. Applying rationality to health choices involves clear, evidence-based thinking and consideration of long-term consequences.

Choosing a Healthy Lifestyle

Deciding to live a healthy lifestyle involves daily choices about what to eat, when and how to exercise, and how to manage stress. A rational approach involves assessing the long-term benefits of a healthy lifestyle against the short-term sacrifices it might require.

Method: Evaluate the impact of different diets or exercise regimes not just on your physical health, but also on your mental health and overall quality of life. Use this analysis to choose a lifestyle that is not only health-promoting but also sustainable and enjoyable for you.

Medical Decisions

When faced with medical decisions, especially under conditions of uncertainty, rational decision-making becomes crucial. This might involve choosing treatments or deciding on preventative measures.

Approach: Gather as much reliable information as possible, consult with medical professionals, and consider both the immediate impact of a decision and its long-term implications on your health and quality of life. For example, when choosing between treatments, consider factors such as side effects, success rates, and how the treatment aligns with your personal values and lifestyle.

Risk Assessment: Evaluating and Mitigating Risks

Risk assessment is about recognizing potential dangers and making choices that minimize the likelihood of adverse outcomes while maximizing opportunities for positive results.

Personal Safety

Making decisions about personal safety, such as whether to engage in a particular sport or activity, requires a careful evaluation of the risks involved versus the enjoyment or benefits gained.

Analysis: Consider not only the statistical likelihood of injury but also how taking certain safety measures (like wearing protective gear) can mitigate these risks. Decide how much risk is acceptable for the associated level of enjoyment or benefit.

Emergency Planning

Rational decision-making is essential when planning for emergencies, such as natural disasters or financial crises. This involves assessing potential threats and preparing in a way that reduces risk.

Strategy: Develop an emergency plan that includes savings to cover unexpected expenses, supplies to deal with natural disasters, and insurance to mitigate financial losses from major accidents or health issues. Regularly review and adjust this plan as your circumstances or the external environment changes.

Career Application: Strategic Career Planning, Responding to Market Changes, Skill Development

Whether you are just starting out, looking to pivot in your career, or aiming to climb higher in your current path, understanding how to strategically plan your career, respond to market changes, and continuously develop your skills is essential. Let's delve into how you can apply rational decision-making to these critical aspects of career management, empowering you to make choices that not only align with your professional aspirations but also enhance your overall job satisfaction and success.

Strategic Career Planning: Charting a Path Forward

Strategic career planning is not just about choosing a job; it's about mapping out a career path that leads to fulfilling your professional and personal goals. This process requires a clear understanding of your aspirations, the opportunities available, and the steps needed to reach your desired end state.

Identifying Career Goals

The first step in strategic career planning is to identify your long-term career goals. What do you want to achieve in your career? Are you aiming for leadership positions, entrepreneurial ventures, or expertise in a particular field? Clarifying these goals provides a target to aim for and a benchmark against which to measure your progress.

Reflection: Consider what success looks like to you in a career. Is it financial stability, recognition in your field, work-life balance, or something else? Define what matters most to you.

Mapping Out Your Career Path

Once you have identified your goals, the next step is to map out a career path that leads to these goals. This involves understanding the industry landscape, identifying key milestones and positions that will advance your career, and recognizing the skills and experiences required to qualify for these roles.

Action: Create a timeline of your career path, noting important milestones such as potential job changes, additional training or education needed, and other professional development activities. This plan should be flexible, allowing for adjustments as you progress and as circumstances change.

Responding to Market Changes: Staying Relevant

In today's fast-paced world, industries and job requirements are constantly evolving. Staying relevant and adaptable is crucial for long-term career success. Rational decision-making can help you respond effectively to these changes by encouraging a proactive approach to learning and adaptation.

Monitoring Industry Trends

Keeping abreast of industry trends and changes in your field is essential. This can involve regular research, attending industry conferences, and engaging with professional networks to stay informed about new technologies, shifts in market demand, and emerging opportunities.

Implementation: Set up a routine for staying updated, such as subscribing to industry newsletters, following thought leaders on social media, and participating in professional groups. Use this information to assess how market changes might impact your career path and what new skills you may need to develop.

Adapting to New Technologies and Methods

As new technologies and methodologies emerge, adapting quickly is a key competitive advantage. This might mean learning new software, acquiring new techniques, or even shifting to a different but burgeoning area of your field.

Strategy: Identify resources for learning new skills, such as online courses, workshops, or mentorship opportunities. Allocate regular time in your schedule for this development, ensuring that your skills remain on the cutting edge.

Skill Development: Continuous Learning and Growth

Continuous skill development is crucial not just for adapting to market changes but also for advancing along your chosen career path. Rational decision-making supports a structured approach to identifying and acquiring the skills needed for future career opportunities.

Assessing Skills Gaps

Regularly assess your current skills against the requirements of your desired career trajectory. This assessment can help identify any gaps that might hinder your progress towards your career goals.

Evaluation: Periodically review job descriptions of roles you aspire to, and compare the required skills and qualifications with your current capabilities. This can highlight areas where further development is needed.

Pursuing Targeted Learning Opportunities

Once you've identified the skills you need to develop, pursue targeted learning opportunities. These might include formal education, on-the-job training, or self-directed learning.

Plan: Choose learning opportunities that offer the best potential for growth in your specific areas of need. Consider the quality of the program, the applicability of the skills taught, and how well it fits into your overall career plan.

Business Application: Rational Strategies in Negotiations, Risk Management, Long-term Planning

Rational Strategies in Negotiations

Negotiations are a cornerstone of business activity, encompassing everything from procurement and sales agreements to partnerships and labor contracts. A rational approach to negotiations involves more than just aiming for the best price; it requires a comprehensive strategy that considers long-term relationships, market conditions, and the broader business objectives.

Preparation and Planning

The first step in rational negotiation is thorough preparation. This involves understanding your own goals and the limits of what you are willing to accept. It also requires a deep understanding of the other party's interests and how much you can flexibly respond to their needs without compromising your own critical objectives.

Example: Picture yourself in the midst of negotiating a contract with a supplier. A rational approach would involve detailed preparation where you not only understand your pricing limits but also consider other factors such as delivery timelines, payment terms, and quality assurances that might be negotiable. You would also research the supplier's recent activities, financial health, and other partnerships to gauge their priorities and pressures.

Strategic Execution

During negotiations, employing a rational strategy means actively managing the dialogue to steer towards a mutually beneficial outcome. This involves clear communication, strategic concessions, and sometimes, the willingness to walk away if the deal doesn't meet your critical business needs.

Approach: Use a combination of open questions and assertive statements to clarify points and assert your position. Be prepared to offer strategic concessions that cost you little but may have high value to the other party, thereby creating goodwill and facilitating a more favorable agreement.

Risk Management: Enhancing Business Resilience

Strong risk management is essential for ensuring a business's stability and profitability. Making informed decisions about risk involves recognizing possible threats, evaluating their probability and impact, and devising effective strategies to minimize their effects.

Identifying and Prioritizing Risks

A fundamental step in effective risk management is carrying out a thorough evaluation of risks. This process includes pinpointing every possible threat that might impact your business, ranging from financial uncertainties and operational setbacks to technological threats and strategic challenges.

Application: Develop a risk register that lists all identified risks, assesses their potential impact on the business, and estimates the probability of their occurrence. This document becomes a crucial tool in prioritizing which risks need immediate attention and resources.

Mitigation and Contingency Planning

Once risks are identified and prioritized, the next step is to develop mitigation strategies. This includes both actions to reduce the likelihood of risks occurring and contingency plans in case they do occur.

Strategy: For instance, if a key risk is supplier failure, mitigation strategies might include diversifying suppliers or developing in-house capacities to ensure continuity. Contingency plans might involve pre-negotiated agreements with alternate suppliers or maintaining a reserve stock of critical supplies.

Long-term Planning: Securing Future Success

Long-term planning in business requires a vision that extends beyond the immediate operational needs. Rational long-term planning involves setting strategic goals based on careful analysis of market trends, internal business capacities, and potential future challenges.

Setting Strategic Goals

The process of setting strategic goals should be grounded in a rational analysis of where the business needs to go to succeed. This involves not only setting financial targets but also considering product development, market expansion, and potential diversification.

Approach: Conduct scenario planning sessions where different future scenarios are imagined and planned for. This helps in understanding how different strategies might play out under various future conditions, allowing you to develop flexible strategies that can adapt to changes.

Implementing Strategic Plans

The implementation of strategic plans requires detailed action plans that outline specific steps, allocate resources, and set timelines. This process also requires regular reviews and updates as market conditions change and new opportunities arise.

Execution: Break down strategic goals into actionable objectives for each department or team. Assign clear responsibilities and deadlines, and establish regular review meetings to assess progress against the strategic plan. Be prepared to adjust plans as feedback and new data come in, ensuring that the business remains agile and responsive to change.

Government Application: Policy-making, Public Administration, Crisis Management

Rational Decision-Making in Policy-Making

Policy-making involves the development of policies intended to address public issues and improve the common welfare. Rational decision-making in this context ensures that policies are not only well-designed to meet their objectives but are also resource-efficient and equitable.

Developing Effective Policies

Effective policy-making requires a thorough analysis of the problem at hand, consideration of various potential solutions, and the selection of an option that offers the most significant benefit to the public. This process involves gathering extensive data, forecasting outcomes, and evaluating the impact of different policy options.

Example: Consider the development of a national healthcare policy. A rational approach would involve analyzing various healthcare models, assessing public health data, considering budget constraints, and predicting the long-term impacts of different healthcare systems on public health and economic stability.

Incorporating Stakeholder Input

Effective policy-making also requires the inclusion of diverse stakeholder perspectives. This ensures that the policies developed are not only comprehensive but also considerate of the needs and values of different community segments.

Engagement: Facilitate forums, surveys, and public consultations to gather input from a broad range of stakeholders, including healthcare providers, patients, insurance companies, and economic experts, to ensure that the policy addresses broad and specific needs effectively.

Rational Decision-Making in Public Administration

Public administration involves managing public agencies, implementing policies, and overseeing the bureaucratic functions of the government. Rational decision-making in this sphere ensures that public resources are managed efficiently and that governmental operations align with the established policies and goals.

Resource Allocation

A key aspect of public administration is the efficient allocation of resources. This involves deciding how to distribute limited resources such as funding, manpower, and materials across various government projects and sectors.

Application: In managing a city's budget, rational decision-making would involve using financial analysis to allocate funds strategically across essential services such as education, public safety, and infrastructure. This requires evaluating the impact of funding allocations on service quality and public satisfaction.

Streamlining Operations

Improving the efficiency of government operations is another crucial aspect of public administration. This involves optimizing processes and procedures to reduce waste, eliminate redundancy, and enhance service delivery.

Implementation: Introduce technology solutions for automating routine tasks, develop performance metrics to evaluate and improve employee productivity, and redesign workflows to streamline service delivery processes.

Rational Decision-Making in Crisis Management

Crisis management involves planning for and responding to emergencies such as natural disasters, economic crises, or public health emergencies. Rational decision-making is crucial here to quickly assess situations, make informed decisions, and deploy resources effectively to mitigate the impact of crises.

Preparedness Planning

Effective crisis management begins with preparedness. This involves developing plans that anticipate various emergency scenarios and outline effective response strategies.

Strategy: Conduct risk assessments to identify potential emergencies and develop comprehensive response plans that include evacuation routes, resource deployment strategies, and communication plans to inform and protect the public during crises.

Responsive Decision-Making

During a crisis, decision-making must be swift and based on the best available information. This involves continuously monitoring the situation, adjusting responses as new information becomes available, and coordinating with various stakeholders to ensure a unified and effective response.

Response: In the event of a natural disaster, quickly assess the areas most impacted and prioritize resource deployment to those areas. Use real-time data to adjust strategies and responses as the situation evolves, ensuring that efforts are focused where they are needed most.

Legal Application: Litigation Strategies, Negotiation in Legal Settings, Legal Advising

Litigation Strategies: Navigating the Courtroom with Rational Tactics

Litigation involves complex decision-making with high stakes. The rational approach in this context ensures that every legal action is calculated to advance towards a favorable outcome, considering both the immediate impacts and the longer-term repercussions.

Developing a Case Strategy

The foundation of effective litigation is a well-formulated case strategy. This involves a thorough analysis of the facts, a clear understanding of the relevant law, and an astute assessment of the opposition's potential arguments.

Approach: Start by gathering as much information as possible about the case. Review documents, interview witnesses, and research legal precedents. Next, identify the strengths and weaknesses of your case. Use this analysis to craft a strategy that emphasizes your strengths and mitigates your weaknesses, considering how each action you take will be perceived by a judge or jury.

Predictive Analysis in Litigation

Utilizing predictive analysis can significantly enhance your litigation strategy. This involves using data about similar past cases to predict outcomes and inform decisions regarding whether to settle or proceed to trial.

Implementation: If you're deciding whether to settle a case or go to court, rational decision-making involves comparing the predicted outcomes based on statistical data with the client's goals and risk tolerance. For example, if data shows that similar cases have a high likelihood of favorable settlements, and the settlement offer on the table aligns with your client's objectives, recommending settlement might be the rational choice.

Negotiation in Legal Settings: Crafting Win-Win Solutions

Negotiation is a pivotal skill in legal practice, used not only in dispute resolution but also in deal-making scenarios. Rational strategies here involve seeking outcomes that satisfy all parties to the extent possible, promoting lasting agreements and preserving relationships.

Understanding Interests and Leverage

In legal negotiations, understanding the underlying interests of all parties involved is crucial. This goes beyond the surface demands to what each party truly needs to feel satisfied with an agreement.

Technique: In a negotiation over a business contract, take the time to understand not just what the other party is asking for but why they are asking for it. This understanding allows you to propose alternative solutions that may satisfy their needs in ways that are more favorable or feasible for your client.

Balancing Assertiveness and Empathy

Effective negotiation requires a balance between assertiveness to advocate for your client's interests and empathy to understand and respect the opposing party's position. This balance helps in crafting solutions that are robust and mutually acceptable.

Strategy: Be clear and firm about your client's needs but also listen actively to the other party. Use the insights gained to propose creative, cooperative solutions that advance your client's interests while addressing the concerns of the other party.

Legal Advising: Guiding Clients with Strategic Insight

Legal advising involves helping clients make informed decisions about their legal matters. Rational decision-making in this area ensures that advice is not only legally accurate but also strategically sound, taking into account the broader context of the client's personal or business objectives.

Analyzing Options and Outcomes

When advising clients, it's important to present a rational analysis of all available options and their potential outcomes. This includes a clear assessment of risks and benefits associated with each course of action.

Application: If advising a client considering a lawsuit, analyze and communicate the potential financial costs, the likelihood of winning, the possible damages recoverable, and the personal or business impacts of prolonged litigation. Help the client weigh these factors against their goals and resources to come to a rational decision.

Future-Oriented Thinking

Legal advice should also be future-oriented, considering how today's decisions will impact the client's future legal standing and broader life or business goals.

Consideration: In estate planning, advise clients on how different arrangements might protect their assets and affect their heirs. Discuss not only the legal implications but also potential familial or business conflicts that might arise, helping clients make decisions that are beneficial and sustainable long-term.

Environmental Application: Policy Design, Resource Allocation, Environmental Risk Management

Whether you are an environmental policy maker, a manager in a sustainability-focused organization, or simply a concerned citizen, understanding how to apply rational strategies in policy design, resource allocation, and environmental risk management can significantly influence the effectiveness of environmental actions and initiatives.

Rational Environmental Policy Design

Designing environmental policies that effectively address complex ecological challenges requires a blend of scientific understanding, economic analysis, and strategic foresight. Rational decision-making in this context ensures that policies are not only scientifically sound but also economically viable and socially acceptable.

Incorporating Scientific Research and Data

Effective environmental policy must be grounded in robust scientific research. Rational policy design begins with a thorough analysis of environmental data and scientific findings to understand the extent and causes of environmental issues.

Example: Consider the design of a policy aimed at reducing air pollution in a major city. The rational approach involves analyzing pollution data to identify the most significant sources of emissions, reviewing scientific studies on the health impacts of pollution, and using this information to formulate regulations that target the most problematic pollutants and sources.

Stakeholder Engagement and Feedback Integration

Rational environmental policy design also requires the engagement of all stakeholders affected by the policy. This includes businesses, local communities, environmental groups, and government agencies. Engaging these stakeholders helps to ensure that the policy is comprehensive and considers the perspectives and needs of all parties involved.

Approach: Conduct stakeholder meetings, public consultations, and surveys to gather input on proposed environmental policies. Use this feedback to refine the policy, ensuring that it addresses the concerns of stakeholders while still achieving its environmental objectives.

Rational Resource Allocation in Environmental Projects

Resource allocation in environmental management involves distributing limited resources such as funding, manpower, and technical expertise in a way that maximizes the environmental benefit.

Prioritizing Projects Based on Impact and Feasibility

Not all environmental projects are equally impactful or feasible. Rational resource allocation requires evaluating projects based on their potential environmental benefits and the likelihood of successful implementation.

Strategy: Develop a scoring system to evaluate environmental projects based on criteria such as potential impact on biodiversity, water quality improvement, carbon reduction, cost-effectiveness, and community support. Allocate resources preferentially to projects that score highest on these criteria, ensuring that limited resources are used in the most effective manner possible.

Balancing Short-term Needs with Long-term Sustainability

In environmental management, it is crucial to balance the urgency of immediate environmental needs with the goal of long-term sustainability. This involves making decisions that may require more significant investment now but promise greater environmental and economic benefits in the future.

Implementation: For instance, investing in renewable energy infrastructure may involve high initial costs but leads to long-term benefits in terms of reduced greenhouse gas emissions and lower energy costs. Rational decision-making supports such investments by focusing on comprehensive cost-benefit analyses that account for long-term environmental and economic impacts.

Rational Environmental Risk Management

Environmental risk management involves identifying, assessing, and mitigating risks associated with environmental harm. This could include risks from natural disasters, industrial accidents, or the unintended consequences of human activities.

Identifying and Assessing Environmental Risks

The first step in rational environmental risk management is to systematically identify potential risks and assess their likelihood and potential impact. This assessment helps prioritize risks and tailor mitigation strategies accordingly.

Example: If a chemical manufacturing plant is located near a river, a rational risk assessment would evaluate the potential for accidental chemical spills, the likelihood of such events, and their possible impact on water quality and ecosystem health. This analysis guides the development of targeted risk mitigation measures, such as improved safety protocols and emergency response plans.

Developing and Implementing Mitigation Strategies

Once risks are assessed, the next step is to develop and implement strategies to mitigate those risks. This involves not only planning and resource allocation but also ongoing monitoring and adaptation of strategies based on new information and changing conditions.

Plan: Implement regular environmental monitoring around the chemical plant to detect any signs of pollution early. Establish a rapid response team equipped and trained to deal with chemical spills, and conduct regular drills to ensure preparedness.

Reflection Questions

1) Evaluating Rationality: Consider a significant decision you recently faced in your personal or professional life. Using the criteria for rationality outlined in this chapter, assess whether your decision-making process was rational. What factors most influenced your level of rationality?

2) Expected Utility Theory in Action: Reflect on a situation where you had to choose between options with uncertain outcomes, such as in financial planning or health choices. How could applying the principles of Expected Utility Theory have helped you make a more informed decision? What limitations of this theory might have impacted your decision?

3) Limitations of Rational Decision-Making: Think about a decision in a business, government, or environmental context that you are familiar with, where rational decision-making seemed to fall short. What were the biggest challenges or limitations faced? How could the decision-making process have been improved by acknowledging these limitations?

Part 2 - Core Game Theory Concepts

Section 4: Nash Equilibrium and Strategic Stability

What is Nash Equilibrium? Definition and Foundational Concepts

The Nash Equilibrium is one of the most pivotal concepts in game theory. A concept that revolutionized the understanding of strategic interactions in economics, politics, business, and beyond. Whether you're a student, a professional, or simply a curious mind, understanding Nash Equilibrium will enhance your ability to analyze situations where the outcome depends not only on your own decisions but also on those of others.

Understanding Nash Equilibrium

Nash Equilibrium, named after mathematician John Nash, who formulated it, is a key concept in the study of game theory that occurs when all participants in a game or economic model choose strategies that maximize their own payoff, given the strategies of others. At this point, no player can benefit by changing their strategy unilaterally. This creates a state of strategic stability, crucial in various decision-making environments.

The Essence of Nash Equilibrium

Imagine you are playing a game where each player must choose a strategy without knowing the choices of the others. Nash Equilibrium occurs when each player's choice becomes the best response to the choices of the others. In essence, once equilibrium is reached, no player has an incentive to deviate from their strategy, assuming others will stick to theirs.

Example: Consider a two-player game where each has to choose between cooperating or betraying the other. Nash Equilibrium in this scenario might be both players betraying each other if betraying provides a higher reward or a lesser penalty regardless of the other player's choice, making betrayal a dominant strategy for both.

The Foundations of Nash Equilibrium

To truly grasp Nash Equilibrium, it's essential to understand some foundational concepts that underpin it. These include best responses, strategy profiles, and the equilibrium's conditions.

Best Response Strategy

A best response is a strategy or action that yields the highest payoff for a player, given the strategies chosen by other players. Identifying best responses is crucial for determining Nash Equilibrium as it involves analyzing how changes in one player's strategy affect the other's payoff.

Application: In business negotiations, a company's best response might be adjusting their offer to match a competitor's unless doing so would result in a net loss. Here, the best response depends critically on the competitor's actions.

Strategy Profiles

A strategy profile includes all the strategies chosen by all players in the game. Nash Equilibrium occurs within a specific strategy profile where all players are playing their best response strategies.

Insight: In a market competition, the strategy profile might involve different firms choosing their production quantities. The Nash Equilibrium would be the set of quantities where each firm is maximizing its profit given the quantities chosen by the others.

Conditions for Nash Equilibrium

For a Nash Equilibrium to exist, certain conditions must be met:

1) Knowledge of Strategies: Each player must know the strategies available to them and predict others' strategies.

2) Rationality: Players are assumed to act rationally, seeking to maximize their own payoffs.

3) No External Influence: The equilibrium assumes no changes or external influences disrupt the players' payoffs from their chosen strategies.

Real-World Implications of Nash Equilibrium

Understanding Nash Equilibrium is not just an academic exercise; it has real-world implications across various fields:

- **Economics**: In markets, firms reach a Nash Equilibrium in pricing and production decisions, balancing competitive pressures against market demands.
- **Politics**: Political parties often reach Nash Equilibria during negotiations, where no party can further its agenda without cooperation or concessions from others.
- **Social Dynamics**: In social settings, individuals reach equilibria in behaviors and norms, especially in situations involving trust and reciprocity.

Reaching Nash Equilibrium: How Players Converge on a Strategy Set

Understanding the Convergence to Nash Equilibrium

Reaching Nash Equilibrium is about understanding how individuals adjust their strategies in response to the choices of others until no one wishes to change their strategy anymore. This process is not just about choosing a strategy in isolation but about anticipating and reacting to the strategies of others.

The Dynamics of Strategy Adjustment

Imagine you're playing a game where the goal is to guess a number closest to two-thirds of the average number guessed by all players. Initially, everyone might guess randomly, but as rounds progress, players will adjust their guesses based on previous outcomes. Each player thinks, "If others are guessing high, I should guess lower to be closer to two-thirds of the average." Over successive rounds, guesses typically converge to zero as each player continuously adjusts, aiming to reach the best response given the others' strategies.

Feedback and Adaptation

In many strategic situations, feedback plays a crucial role in reaching Nash Equilibrium. Feedback can be direct, such as explicit responses from other players, or indirect, such as observing outcomes and adjusting accordingly.

Example: In a business context, consider a company pricing its product. It initially sets a price based on expected competitor prices. Once the product is launched, the company might receive feedback in the form of sales volumes and market share. If the product is selling less than expected, the company might lower the price. Conversely, if the product sells more than anticipated, the price might be increased. This process of adjusting pricing strategies continues until the company finds a price point where it no longer feels necessary to adjust—essentially reaching a Nash Equilibrium.

Factors Influencing Convergence to Nash Equilibrium

Several factors can affect how and whether players converge to a Nash Equilibrium. Understanding these factors can help you predict and influence the process in real-world settings.

Transparency and Information Availability

The availability of information about other players' actions significantly impacts the convergence process. When more information is available, players can make more informed decisions about their strategies.

Scenario: In financial markets, traders make buying or selling decisions based on information about market conditions and other traders' actions. More transparent markets, where information is readily available, tend to reach equilibrium faster because traders can quickly adjust their strategies based on comprehensive market data.

Complexity of the Game

The complexity of the game or scenario also affects how easily Nash Equilibrium is reached. In simple games with a few players and clear choices, reaching equilibrium can be straightforward. However, in complex games with many players and strategic options, convergence can be more challenging and require more iterations of strategy adjustments.

Illustration: Consider a large corporation with multiple divisions competing for a limited budget. Each division might initially request more funds than it needs, anticipating cuts. Over time, as each division sees the outcomes of their requests—whether they receive more or less funding than requested—they adjust future requests. Achieving equilibrium, where each division requests exactly what it needs without further adjustments, might take several budget cycles.

The Role of Rationality in Reaching Equilibrium

Rationality is a foundational assumption in the theory of Nash Equilibrium. It assumes that all players are rational, meaning they consistently act in a way that they believe will lead to the best possible outcome for themselves based on the information available.

Consistent Strategy Adjustments

For equilibrium to be reached, players must not only be rational but also consistent in their strategy adjustments. This consistency is crucial in stabilizing the game or scenario, as it allows players to predict each other's behavior and adjust their strategies accordingly.

Example: In contract negotiations between a union and management, both sides present demands and concessions. If both sides are rational and consistent in their negotiations, recognizing the constraints and needs of the other side, they are more likely to adjust their demands to reach an agreement that resembles Nash Equilibrium, where neither side has an incentive to deviate from the agreed terms.

Examples of Nash Equilibrium: Simplified Illustrations Across Various Fields

These examples will help you see how Nash Equilibrium isn't just a theoretical concept but a real-world phenomenon that affects decision-making and strategy in multiple facets of life. By understanding these examples, you can better appreciate how strategic stability and mutual best responses play out in everyday situations.

Nash Equilibrium in Economics

In economics, Nash Equilibrium frequently appears in market dynamics where companies decide on pricing, production levels, or entering new markets.

Example: Duopoly Market

Imagine two companies, Company A and Company B, selling a similar product. If Company A decides to lower its prices to increase its market share, Company B might react by lowering its prices as well. However, if both companies continue lowering their prices, they will eventually reach a point where neither can afford to lower prices further without incurring losses. At this point, both companies decide it's best not to change their pricing strategy despite the competitive pressure. This price point represents a Nash Equilibrium because neither company benefits from unilaterally changing its prices, given the current strategy of the other.

Nash Equilibrium in Politics

In politics, Nash Equilibrium can explain the stability of certain policies or the deadlock in negotiations.

Example: Legislative Deadlock

Imagine several political parties are trying to form a government together. Each party wants to have as much influence as possible, but they also need to make some compromises to be part of the government. A Nash Equilibrium is reached when they all agree on how to share power in a way that no single party would benefit by leaving the group unless doing so would cause the government to collapse. This situation helps explain why some government coalitions can remain stable even though the parties involved may have very different goals and ideas.

Nash Equilibrium in Personal Life

Nash Equilibrium also appears in personal decision-making, especially in scenarios involving family or household decisions.

Example: Household Chores

Imagine a couple deciding who does which household chore. Each person prefers the other to do the more tedious tasks. However, they might reach an arrangement where each person takes on certain responsibilities that balance the overall workload. This distribution becomes a Nash Equilibrium when neither partner would benefit by changing their individual set of chores unless the other also makes a change, as this would disrupt the balance and increase conflict.

Nash Equilibrium in Career

In career-related decisions, Nash Equilibrium can illustrate scenarios where job market competitors or coworkers reach a stable arrangement.

Example: Job Promotion Competition

Suppose several employees are competing for a promotion. Each has the option to work extra hours to outperform the others. However, if all competitors decide to work significantly longer hours, they might end up equally likely to get the promotion, but all at a higher personal cost (less personal time, increased stress). They might reach a Nash Equilibrium where each person decides it's not worth working additional hours beyond a certain point because the extra effort doesn't increase their likelihood of promotion enough to justify the cost.

Nash Equilibrium in Business

In business strategy, companies often reach Nash Equilibria in competitive tactics and partnership decisions.

Example: Advertising Spend in a Competitive Market

Two businesses competing in the same market might decide on their advertising spends. If each company increases its advertising budget, the overall market share might not change significantly, but both incur higher costs. A Nash Equilibrium is reached when both decide to maintain a certain level of advertising that maximizes return on investment without unnecessarily escalating costs.

Nash Equilibrium in Technology

In technology, especially in software and network settings, Nash Equilibrium can explain choices regarding system compatibility and user behavior.

Example: Platform Adoption

Users of a technology platform or software might face a choice between competing technologies. A Nash Equilibrium occurs when users settle on a platform because most others are using it, making it de facto the standard, even if other platforms might offer superior features.

Nash Equilibrium in Social Settings

Social interactions often lead to Nash Equilibria, where individuals adjust their behavior based on social norms and the expected behavior of others.

Example: Social Conformity

In social groups, individuals often conform to the group norms to avoid standing out negatively. A Nash Equilibrium is reached when everyone in the group conforms to these norms because deviating from them would lead to social sanctions or disapproval, even though expressing individuality might be personally preferable.

Reflection Questions

1) Identifying Nash Equilibrium: Think of a competitive situation you have observed or participated in, perhaps in a business context or a personal scenario. Try to identify if and how a Nash Equilibrium might have been reached. What were the strategies involved, and how did they contribute to creating stability in the situation?

2) Challenges in Reaching Nash Equilibrium: Reflect on a decision-making process involving multiple stakeholders (like in a corporate setting or family decision). What challenges were faced in converging on a mutually agreeable strategy? Discuss any real or perceived barriers to reaching a Nash Equilibrium.

3) Practical Applications: From the examples given in this section, choose one that you find most relevant to your life. Explain how understanding Nash Equilibrium can help better analyze the situation and predict the outcomes of different strategic choices. How might this knowledge influence your approach if you were one of the players in this scenario?

Section 5: The Prisoner's Dilemma & Other Example "Games"

Definition and Explanation: The Prisoner's Dilemma – A Classic Example of Game Theory

The Prisoner's Dilemma is one of the most famous and illuminating concepts in game theory. This classic example not only provides foundational insights into the nature of strategic interactions but also helps us understand the complexities of decision-making in situations where individuals' interests are both conflicting and complementary. By dissecting this dilemma, you'll gain a profound understanding of the theoretical underpinnings that influence a wide range of real-world scenarios across various fields.

Understanding the Prisoner's Dilemma

The Prisoner's Dilemma is a standard example used in game theory to demonstrate why two completely rational individuals might not cooperate, even if it appears that it is in their best interests to do so. This dilemma is structured around a hypothetical situation involving two prisoners suspected of committing a crime together. Here's how it typically unfolds:

The Setup

Two criminals are arrested and imprisoned. Each prisoner is held in solitary confinement, unable to communicate with the other in any way. The prosecutors do not have enough evidence to secure a conviction on the main charge against the pair. They aim to have both prisoners sentenced to a year in prison on a lesser charge. At the same time, the prosecutors present each prisoner with a deal. Each one can choose either to betray the other by accusing them of committing the crime, or to collaborate by choosing to remain silent. The possible outcomes are:

- **Mutual Cooperation (Both Refuse to Confess)**: If both prisoners decide to remain silent, they both serve only one year in prison (on the lesser charge).
- **Mutual Defection (Both Confess)**: If both prisoners betray each other, they both serve two years in prison.
- **One Cooperates (Refuses to Confess), One Defects (Confess)**: If one prisoner betrays the other, the one who betrays will be set free, while the other serves three years in prison.

The Dilemma and How to Solve It

Rationally, the best outcome for each prisoner individually occurs if they betray the other. If each follows a purely individual rational decision strategy, both will likely betray the other, leading them to both serve two years in prison — a worse outcome than if both had cooperated by remaining silent.

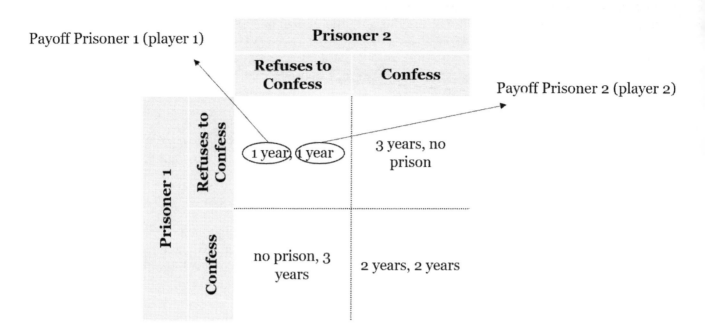

The illustration above shows the situation and the payoff for each prisoner (player). For example, if Prisoner 1 refuses to confess he will get 1 year of prison and the same is for prisoner 2. If prisoner 1 confesses and prisoner 2 refuses to confess then prisoner 1 will get no prison time while prisoner 2 will get 3 years of prison and vice versa. If they both confess they both get 2 years of prison. They way we solve this is to look from the point of view of each player.

Player 1 – Looking at it rationally, if prisoner-1 assumes that prisoner-2 refuses to confess then player-1 best strategy is to confess and get no prison time instead of 1 year in case of refusing to confess. If prisoner-1 assumes that prisoner-2 will confess also in this case prisoner-1 best strategy is to confess as he will get 2 years instead of 3 years. Therefore for prisoner-1 the best is to confess.

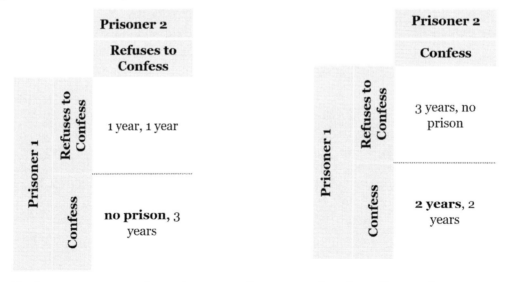

Player 2 – if prisoner-2 assumes that prisoner-1 refuses to confess then player-2 best strategy is to confess also and get no prison time instead of 1 year in case of refusing to confess. If prisoner-2 assumes that prisoner-

1 will confess, also in this case prisoner-2 best strategy is to confess as he will get 2 years instead of 3 years. Therefore for prisoner-2 the best strategy is to also to confess

		Prisoner 2	
		Refuses to Confess	Confess
Prisoner 1	Refuses to Confess	1 year, 1 year	3 years, **no prison**

		Prisoner 2	
		Refuses to Confess	Confess
Prisoner 1	Confess	no prison, 3 years	→**2 years, 2 years**

The Final Outcome – for both of them following this reasoning the best strategy is to confess. The solution to the Prisoner's Dilemma is that both prisoners confess and get 2 years each.

		Prisoner 2	
		Refuses to Confess	Confess
Prisoner 1	Refuses to Confess	1 year, 1 year	3 years, no prison
	Confess	no prison, 3 years	**2 years, 2 years**

Key Concepts Illustrated by the Prisoner's Dilemma

Nash Equilibrium

In the Prisoner's Dilemma, the Nash Equilibrium occurs when both prisoners choose to defect (betray). Despite the fact that mutual cooperation yields a better collective outcome, the Nash Equilibrium demonstrates the individual rationality leading to a collectively worse outcome.

Dominant Strategy

Defection (betraying the other) is considered a dominant strategy in the traditional Prisoner's Dilemma because it produces a better outcome for a player regardless of the other player's decision. This is because betraying a cooperating partner yields the best payoff (freedom), and betraying a betraying partner avoids the worst payoff (three years in prison).

Pareto Optimality

The concept of Pareto Optimality is also important in understanding the Prisoner's Dilemma. The mutual cooperation outcome is Pareto optimal because moving away from this outcome would make at least one player worse off without making the other player better off.

The Broader Implications of the Prisoner's Dilemma

The Prisoner's Dilemma is more than just a theoretical construct; it has real-world implications across various domains. It illustrates the often counterintuitive nature of strategic decision-making and highlights the challenges of achieving optimal outcomes when individual incentives do not align with collective welfare. This dilemma is particularly relevant in understanding economic competitions, political negotiations, environmental agreements, and any situation where the interplay between cooperative and selfish behaviors influences the outcome.

Real-World Examples: Situations of the Prisoner's Dilemma in Various Fields

The Prisoner's Dilemma finds practical application in a vast array of scenarios beyond the theoretical. These situations span economics, politics, personal life, career, business, sports, technology, social contexts, environmental issues, and legal affairs. Understanding how the dynamics of the Prisoner's Dilemma play out in these real-world contexts not only enriches our understanding of strategic decision-making but also enhances our ability to navigate complex interactions. This extensive exploration will delve into each of these fields, illustrating how the principles of the Prisoner's Dilemma influence behaviors and outcomes in diverse settings.

Economics: The Dilemma in Market Competition

In economics, the Prisoner's Dilemma frequently appears in competitive markets where businesses must decide between cooperative and competitive strategies.

Example: Price Wars Among Competitors

Consider two competing companies in the same industry. If both companies decide to lower prices to outcompete each other, they end up in a price war, which can erode profits for both. However, if they both keep prices high, they maintain higher profit margins. Despite the mutual benefits of maintaining high prices, the temptation to undercut the competitor to gain market share often leads both to lower prices, resulting in reduced profits for both parties. This scenario is a classic representation of the Prisoner's Dilemma, where rational individual strategies lead to a worse collective outcome.

Politics: Bipartisan Cooperation

In politics, the Prisoner's Dilemma manifests in situations where parties must decide between cooperation and opposition.

Example: Legislative Deadlocks

Political parties in a legislature may benefit from cooperating on certain bipartisan laws that would be universally beneficial, such as infrastructure improvements. However, the fear that cooperating (or appearing to cooperate) with the opposition might damage a party's standing with its base often leads both parties to oppose each other. As a result, legislation stalls, and no one achieves their objectives, illustrating the dilemma of individual rationality leading to collective irrationality.

Personal Life: Relationship Decisions

The dynamics of the Prisoner's Dilemma are also evident in personal relationships, particularly in conflict resolution.

Example: Household Responsibilities

In a household, two partners might each prefer the other to take on more chores. If both decide to shirk responsibilities, the household chores do not get done, leading to a messy living environment. Cooperation by sharing the workload evenly would clearly be beneficial to both, but the temptation to minimize personal effort can lead to mutual defection, with negative outcomes for the relationship.

Career: Workplace Collaboration

In career settings, the Prisoner's Dilemma can impact collaboration and competition among coworkers.

Example: Sharing Resources

Imagine a scenario where two employees have access to a limited set of resources that could help them excel in their projects. They can either agree to share the resources equitably, ensuring both can progress reasonably well, or they can each attempt to monopolize the resources for their own projects. If both choose to compete for the resources, they might hinder each other's progress, which is detrimental to both their projects.

Business: Strategic Alliances

In the business world, companies often face the Prisoner's Dilemma when deciding whether to form strategic alliances or compete fiercely.

Example: Technology Development

Two tech companies might benefit from pooling their research and development efforts to innovate a new technology. Cooperation would likely lead to better outcomes for both, but the fear of giving a competitive advantage to the other might lead both to work in isolation, potentially duplicating efforts and slowing down innovation.

Sports: Team Dynamics

Sports teams often encounter the Prisoner's Dilemma when individual players decide between personal performance and team strategy.

Example: Passing the Ball

In basketball, a player might face the choice of passing the ball to a teammate for a better shot or taking a difficult shot themselves. While passing might increase the team's chance of scoring (benefiting everyone), the temptation to increase personal scores can lead players to make suboptimal decisions, affecting team performance.

Technology: Data Sharing

In the realm of technology, companies grapple with the decision to share or withhold information that could lead to broader industry advancements.

Example: Cybersecurity Threats

Tech companies might benefit from sharing information about cybersecurity threats to better protect the entire industry. However, companies often hesitate to share such information, fearing it might expose their own vulnerabilities or diminish their competitive edge.

Social: Community Projects

In social settings, community projects often reflect the dynamics of the Prisoner's Dilemma, especially in voluntary participation.

Example: Community Clean-Up

If all community members participate in a clean-up event, everyone benefits from a cleaner, more pleasant environment. However, individuals might decide not to participate, hoping to still benefit from others' efforts without contributing themselves. If too many opt out, the event fails to achieve its goals, detrimentally affecting the whole community.

Environmental: Conservation Efforts

Environmental conservation efforts are frequently hindered by the Prisoner's Dilemma, particularly in shared resource management.

Example: Fishing Quotas

Fishermen sharing a common body of water face the dilemma of adhering to fishing quotas to ensure sustainable fish populations or overfishing to maximize immediate gains. If all fishermen cooperate by following quotas, fish populations remain sustainable for future benefits. However, the temptation to catch more than the quota can lead to overfishing, depleting the resource to the detriment of all.

Legal: Settlement Negotiations

In legal contexts, the Prisoner's Dilemma often arises during settlement negotiations, where parties must decide between cooperation for a mutually agreeable settlement or aggressive litigation for the possibility of a more favorable unilateral outcome.

Example: Business Litigation

Two businesses in a legal dispute over a contract may either choose to settle the dispute through compromise or pursue prolonged litigation. Settlement would save both parties time and legal fees, but the desire to potentially win more through litigation can lead both to choose a costly court battle.

Additional Example of "Games"

These additional example games provide a comprehensive toolkit for analyzing social dilemmas and strategic decision-making across a broad spectrum of real-world scenarios. Each game highlights different aspects of human behavior in strategic contexts, providing valuable insights into cooperation, competition, risk, and reward dynamics. Whether used individually or in combination, these game theoretical frameworks allow us to predict and influence outcomes in various social, economic, and organizational environments. Let's review one by one.

The Ultimatum

The "ultimatum game" is a fascinating study in economic game theory that tests human behavior around fairness, altruism, and spite. This game features two participants: the proposer and the responder. The proposer is given a sum of money—say, $100—and has the task of proposing how to divide this sum between themselves and the responder. The responder then makes a decision: accept the offer, in which case the money is split as proposed, or reject the offer, in which case both players receive nothing.

At its core, the ultimatum game explores the boundaries of self-interest and equitable behavior. According to standard economic theory, which assumes that individuals act purely out of self-interest, the responder should accept any amount of money offered that is greater than zero, as receiving something is better than receiving nothing. However, the results from numerous repetitions of the game across different cultures and demographics reveal a more complex interplay of motives. Typically, offers below 30% of the total amount are often rejected by responders. This outcome suggests that responders are willing to forego their own potential gains to punish the proposer for what they perceive as an unfair offer.

The frequent rejection of low offers highlights the human preference for fairness and indicates that people are concerned not only with their own welfare but also with equitable outcomes. This behavior is seen as a form of altruistic punishment, where the responder punishes the proposer for deviating from fair behavior, potentially as a means to enforce a norm of fairness in broader social contexts.

The ultimatum game also provides insight into strategic behavior. Proposers, knowing that low offers might be rejected, are often motivated to make fairer offers to avoid coming away with nothing. This aspect of the game illustrates the concept of subgame perfect equilibrium in game theory, where the proposer, anticipating the possible reactions of the responder, chooses an offer that is likely to be accepted.

Moreover, variations of the ultimatum game have been employed to explore how factors like group identity, information asymmetry, and previous interactions influence decision-making. For example, when proposers and responders know each other or expect to interact in the future, offers tend to be more generous, and rejections less frequent.

Overall, the ultimatum game challenges the traditional economic view of rational self-interest and significantly contributes to our understanding of social preferences. It shows that people value fairness and are willing to enforce it even at a cost to themselves, thereby integrating social and emotional factors into the analysis of economic decision-making.

Public Goods Game

The "public goods game" is a standard experiment in economics and game theory that explores how individuals contribute to a common pool to achieve a collective benefit, despite facing individual incentives to free ride or under-contribute. This game encapsulates the conflict between personal interests and the common good, a classic example of a social dilemma.

In the public goods game, each participant in a group is given an initial endowment, usually in the form of tokens or money. Participants must decide how much of their endowment to invest in a public pot and how much to keep for themselves. The total amount collected in the public pot is then multiplied by a factor greater than one (to simulate the increased value of collective investment) and evenly distributed among all players, regardless of their individual contribution. The key dilemma for each player is whether to contribute to the public pot and enhance the group's total benefit or to withhold contributions, keep more of their initial endowment, and free ride on the contributions of others.

Empirical results from numerous iterations of the public goods game reveal that while some individuals do contribute generously, contributions typically decrease over successive rounds of the game as players observe others free riding and adjust their behavior accordingly. This decline in contributions illustrates a central problem in many real-world public goods scenarios: the temptation to free ride can lead to suboptimal outcomes for the group as a whole.

Several factors can influence how much players choose to contribute:

- **Group Size**: Larger groups often see lower per capita contributions because the impact of any one person's contribution is diluted, increasing the temptation to free ride.
- **Communication**: Allowing players to communicate before making their contributions tends to increase the contribution levels, as players can make commitments and potentially exert social pressure on each other.
- **Punishment and Reward**: Introducing mechanisms for punishing free riders or rewarding contributors can significantly affect outcomes. Studies show that when players can punish non-contributors, even at a cost to themselves, overall contributions increase.
- **Repetition and Reputation**: In repeated public goods games, where players interact over multiple rounds, reputation becomes important. Players are more likely to contribute if they know their past actions will be known to others in future rounds, which can foster cooperation.

The public goods game has profound implications for understanding and managing real-world public goods like clean air, national defense, public health measures, and any scenario where individual contributions to a collective resource are voluntary. It helps policymakers and scholars understand the conditions under which cooperation is likely to arise naturally and the types of interventions (such as taxation, subsidies, regulatory frameworks) that can help sustain public resources efficiently.

The Dictator Game

The Dictator Game is a simple economic experiment designed to measure the willingness of individuals to share resources unilaterally. It involves two players: the "dictator" and the recipient. The dictator is given a certain amount of money (or tokens) and has the authority to decide how to allocate this amount between themselves and the recipient. The recipient has no power in this situation and must accept whatever the dictator decides to give, which can range from nothing to the entire amount.

The Dictator Game tests theories of altruism and fairness. Unlike the Ultimatum Game, where the recipient can reject the offer (thus potentially leaving both players with nothing), the recipient in the Dictator Game has no leverage. Traditional economic theory, based on rational self-interest, predicts that dictators will give nothing since they can keep all the money without fear of rejection. However, empirical results often show that many dictators do choose to share a portion of the money, indicating the presence of altruism or a concern for social norms and fairness.

This game is particularly useful in social science research to study behavior that deviates from purely selfish economic rationality. It provides insights into how individuals might act in real-life situations where they have power over resources, highlighting the influence of intrinsic fairness and social preferences.

Critics of the Dictator Game argue that the behavior it elicits may still be influenced by external factors like perceived social expectations or the experimental setting itself. Variants of the game include allowing the dictator to take money away from the recipient or adding a charitable organization as a potential recipient to explore different aspects of altruistic behavior.

The Trust Game (Investment Game)

The Trust Game, also known as the Investment Game, examines trust and reciprocity in a controlled setting. It involves two players: the trustor (or sender) and the trustee (or receiver). The trustor is initially given a sum of money and must decide how much to send to the trustee. Any amount sent is multiplied (typically tripled) by the experimenter to increase the stakes. The trustee then decides how much of the multiplied amount to send back to the trustor.

This game is crucial for understanding the dynamics of trust and reciprocal behavior. The trustor's decision to send money is a measure of trust in the trustee, while the trustee's decision to return money tests their inclination towards reciprocity and fairness. Results often show that trust and reciprocity vary widely among individuals and can be influenced by numerous factors, including past experiences, expectations of future interactions, and cultural norms.

The Trust Game is widely used in both economics and psychology to explore and quantify the levels of trust and cooperation in different populations. It has practical implications for designing business contracts, managing teams, and understanding consumer behavior in economic transactions.

Some variations include changes in the information available to the players (e.g., anonymity, repeated interactions), the multiplication factor, and the stakes involved. These modifications help researchers isolate specific factors affecting trust and reciprocity.

The Chicken Game

The Chicken Game is used to model conflict and risk-taking behavior. Two players move towards each other on a collision course; each must choose either to swerve or to continue straight. If one player swerves while the other continues straight, the swerver "loses" and is considered a coward, while the other player "wins" by standing firm. If both players swerve, the outcome is neutral. If neither swerves, both players face a disastrous outcome (a crash).

This game highlights the consequences of aggressive or risk-taking strategies when mutual destruction is a possible outcome. It is particularly relevant in scenarios involving brinkmanship, where showing weakness can be strategically disadvantageous.

The Chicken Game is often applied in international relations to analyze nuclear deterrence, diplomatic negotiations, and military confrontations. It also appears in business scenarios involving competition strategies where firms face off in high-stakes environments, like price wars or market entry decisions.

Variations of the Chicken Game can include different payoffs to simulate real-world incentives and consequences more accurately, or repeated rounds where players may learn and adapt their strategies based on previous outcomes.

These detailed analyses of each game illuminate the complexities and varied applications of game theory in understanding human behavior in strategic interactions.

The Stag Hunt

The Stag Hunt is a game that explores the conflict between safety and social cooperation. It's based on a scenario where two hunters must choose between individually hunting a hare or cooperatively hunting a stag. Hunting the stag requires mutual cooperation as it's too difficult to capture by a single hunter, but it offers a much larger reward. If one hunter goes for the stag while the other goes for the hare, the stag hunter gets nothing and the hare hunter gets a small but guaranteed reward.

The Stag Hunt illustrates the dilemma faced when individuals must decide between a risky cooperative act that leads to higher collective benefits and a safer, selfish act that guarantees a lesser but certain individual benefit. Trust and assurance of mutual cooperation are crucial in this game. The outcomes show how social arrangements might stabilize on different equilibria based on trust and historical precedent—either cooperation on the high-reward activity or defection to the lower-reward but safer option.

This game is used to model social cooperation involving significant trust and risk, such as community efforts, team projects, and large-scale collaborations. It is particularly useful in studying the conditions under which cooperative strategies emerge and are maintained within groups.

Variations of the Stag Hunt include dynamic versions where players have multiple rounds to influence each other's trust levels or where communication between players is allowed to negotiate commitments to cooperate. These variants help explore how communication and experience affect cooperative outcomes.

The Volunteer's Dilemma

The Volunteer's Dilemma addresses a situation where a group benefits from some costly action, but each individual hopes others will incur the cost. Typically, a single volunteer is needed to produce a public good that benefits all. Each player must decide whether to volunteer, incurring a personal cost, or to free ride, hoping others will volunteer.

This game exemplifies the challenges of public goods where the provision of the good is essential but costly, and where the temptation to shirk responsibility can lead to suboptimal outcomes for the group. If no one volunteers, everyone is worse off, illustrating the potential for collective action failure even when benefits exceed costs.

The Volunteer's Dilemma is applicable in scenarios where collective action is needed but individual incentives to participate are weak. It is relevant in emergency responses, organizational management for initiating projects, or community services where the effort of a few can benefit many.

Various modifications of this game explore different group sizes, costs of volunteering, and benefits of the public good to see how these factors influence the willingness to volunteer. Experimental variations sometimes include communication options or reputation systems to study how these factors can motivate volunteering.

The Assurance Game (Coordination Game)

The Assurance Game, also known as the Coordination Game, centers on the challenge of achieving mutual cooperation when each party's payoff is contingent on the cooperation of others. It highlights situations where participants must coordinate their actions for mutual benefit but face uncertainty about each other's intentions.

The Assurance Game typically features two Nash equilibria: both cooperate or both defect. Successful cooperation leads to a high payoff for both players, reflecting a situation where trust and assurance are established. However, if one cooperates and the other defects, the cooperator receives a lower payoff, leading to a risk of defection if players are uncertain about each other's actions.

Consider two companies that could greatly benefit from partnering on a project (e.g., joint research and development). Each company faces the decision to invest resources in the partnership. The best outcome occurs if both invest, but if one invests and the other doesn't, the investor suffers a loss. The dilemma lies in ensuring mutual investment without the certainty of the other's commitment.

This game is particularly relevant in scenarios involving public goods and team projects, where the collective benefit depends on the participation of all parties. It's also applicable in environmental agreements, collaborative business ventures, and any situation where success depends on widespread cooperative behavior.

The Assurance Game underscores the importance of establishing reliable commitments and trust among parties. It shows that when assurance mechanisms (like contracts, credible commitments, or reputation systems) are in place, they can significantly enhance the likelihood of mutual cooperation. The game also highlights the role of trust-building and communication in overcoming coordination challenges.

The Battle of the Sexes

The Battle of the Sexes is a classic game in game theory that illustrates a coordination problem with conflicting preferences. It involves two players, often modeled as a couple, who must decide on an activity to do together. While both players prefer spending time together over being apart, they have different preferences for the activities available. For example, one may prefer going to a football game, while the other prefers attending a ballet.

The game typically presents two Nash equilibria in pure strategies, where either both go to the football game or both go to the ballet. These outcomes represent the scenarios where both players coordinate on one activity or the other. There is also often a mixed-strategy Nash equilibrium where each player randomizes between the two activities to balance their own preferences with the benefit of being together.

Imagine Alex and Sam are planning their date night. Alex loves football, while Sam is passionate about ballet. If they go to the football game, Alex gets greater satisfaction, but Sam gets less (and vice versa for the ballet). However, both would be unhappy if they chose different activities and ended up spending the evening apart. The strategic challenge is finding an agreement that maximizes their shared happiness, even if it means compromising on individual preferences.

This game is useful in exploring how individuals or groups manage to coordinate in situations with conflicting interests. It's relevant in various social, business, and diplomatic contexts where parties need to find a mutually acceptable solution to benefit from cooperation. The game also illustrates the importance of compromise, negotiation, and communication in achieving successful coordination.

The Battle of the Sexes teaches that while conflict in preferences exists, successful outcomes hinge on the willingness to compromise and the effectiveness of communication between parties. It also shows how establishing conventions or fair decision-making processes (like alternating choices) can help resolve such conflicts.

The Centipede Game

The Centipede Game is a complex strategic interaction in an extensive form, which demonstrates how rational decision-making can diverge from actual human behavior. In this game, two players alternately decide whether to continue the game or take the pot of money, which increases with each move. The game is designed so that the earlier a player chooses to take the pot, the less they get compared to if they had waited longer, but waiting increases the risk that the other player will end the game first.

The game starts with a relatively small sum of money and increases as players pass the decision to the next player. Each player has the option, on their turn, to "stop" the game and take the current sum, splitting it according to a predetermined rule, or to "continue," which increases the potential future payoff but risks losing to the other player who might stop the game on their turn. Rational backward induction in game theory suggests that the first player should stop the game immediately to secure a sure small sum, assuming the second player would also act rationally. However, experiments often show that players let the game continue much longer, suggesting that factors like trust, hope, and fairness play significant roles.

Imagine Alice and Bob playing a Centipede Game starting with $2, set to double each turn if continued. Rationality suggests Alice should take the $2 on her first turn, as continuing risks Bob taking the pot on his turn. Despite this, in reality, players might allow the pot to grow, reflecting a trust or expectation that the other player might also allow it to grow further.

The Centipede Game is used in economic and psychological studies to understand decision-making under uncertainty and the impacts of reciprocal altruism and trust. It challenges the purely rational models of economic behavior by showing how real decisions can be influenced by less quantifiable human factors.

This game highlights the discrepancies between theoretical rational behavior and actual human decisions, emphasizing the importance of psychological factors in strategic interactions. It serves as a powerful reminder that economic and strategic models must account for human emotions and social norms to accurately predict behavior.

The Minority Game

The Minority Game is a repeated game involving an odd number of players who must independently choose between two options, such as two different markets or resources. The winners are those who find themselves in the minority group, which dynamically changes as players adjust their strategies based on previous outcomes.

Each player makes a decision aiming to be part of a minority group, with the game's structure leading to complex adaptive behaviors as players learn and anticipate the choices of others. The game reflects real-world situations where success depends on avoiding the crowded choices of the majority, such as in financial markets or managing traffic flows.

Imagine a scenario where traders choose between two fluctuating stocks each day. Each trader wins by choosing the less popular stock of the day. Traders must predict not only the stocks' movements but also the choices of other traders to consistently be in the minority, making profits.

The Minority Game is particularly relevant in financial markets, traffic systems, and any competitive environment where success hinges on avoiding over-saturated options. It is used to study market dynamics, resource allocation, and the emergence of complex patterns from simple decision rules.

This game demonstrates how individual incentives to diverge from the majority can lead to complex system dynamics. It highlights the importance of pattern recognition, predictive abilities, and adaptive strategies in environments where being different is advantageous. The Minority Game also illustrates how collective behavior can emerge from individual decisions, offering insights into the balance between individual actions and collective outcomes.

The El Farol Bar Problem

The El Farol Bar Problem, named after a bar in Santa Fe, New Mexico, is a problem of resource utilization based on limited capacity. In this scenario, patrons decide weekly whether to go to the bar based on their expectation of the crowd size. The ideal situation for each patron is that the bar is not too crowded but not empty either. If more than 60% of the potential patrons decide to go, the bar becomes overly crowded, and the evening is less enjoyable.

Each patron makes their decision independently, based on their predictions of how crowded the bar will be. These predictions are typically influenced by past experiences (historical data) but are made without communication between patrons. The game illustrates a complex system where individual utility maximization does not necessarily lead to collective utility maximization, similar to congestion in traffic flows or bandwidth in communication networks.

If each patron decides to go to the bar when they predict attendance will be below 60%, the outcome heavily depends on the accuracy and diversity of their predictions. If everyone uses the same or similar predictive model, they might all decide to go or stay home at the same time, resulting in suboptimal outcomes (either overcrowding or underuse).

The El Farol Bar Problem is applicable to situations involving congestion and coordination problems without centralized control, like managing traffic, scheduling use of shared resources, and even in financial markets where traders choose times to buy or sell based on expected market volume.

This problem demonstrates the necessity for diverse strategies and decision frameworks in systems where participants independently affect the outcome. It highlights the challenges of prediction in complex adaptive systems and the potential for what is known as the "minority game" dynamics where success comes from being in the minority.

Reflection Questions

1) Personal Connection: Can you think of a time when you faced a dilemma similar to the Prisoner's Dilemma in your personal or professional life? Describe the situation and analyze your decision-making process. Would you have changed your strategy if you had known about the Prisoner's Dilemma beforehand?

2) Real-World Applications: Choose one of the real-world examples of the Prisoner's Dilemma provided in this section. Discuss how the outcome of the situation could have been different if the parties involved had chosen to cooperate rather than compete. What factors might encourage cooperation in such scenarios?

3) Exploring Solutions: The Prisoner's Dilemma typically shows why two completely rational individuals might not cooperate, even if it seems that it is in their best interests to do so. What strategies or changes in the game's structure could be introduced to lead rational players to a cooperative outcome more often?

Section 6: Zero-Sum and Non-Zero-Sum Games

Defining Zero-Sum Games: The Total Sum of Benefits and Losses Equals Zero

Zero-sum games are a fundamental concept in game theory that shapes understanding of various competitive scenarios across different fields. Zero-sum games represent situations where one participant's gain or loss is exactly balanced by the losses or gains of other participants. This means that the total sum of benefits and losses in the game adds up to zero, creating a scenario where for one to win, another must lose. Understanding this concept will enable you to recognize zero-sum dynamics in various real-life situations and effectively strategize within them.

What Are Zero-Sum Games?

Zero-sum games are scenarios in which the interests of participants are strictly opposed, and the total value created by the outcome of the game is fixed. In simpler terms, the amount of "win" available is constant; any gain by one player directly results from a loss by another.

Characteristics of Zero-Sum Games

- **Fixed Pie**: The idea of a "fixed pie" is crucial in zero-sum games. It means there is a set amount of resources or value to be distributed between the players. No additional value can be created or added through cooperation or other means.
- **Win-Lose Situations**: Every zero-sum game is inherently a win-lose situation, where the success of one participant comes at the expense of another. This creates a purely competitive environment.
- **Equality of Payoffs**: The sum of gains and losses among all players equals zero. This balance means that the sum total of all outcomes for all players will always be zero, marking the zero-sum nature of the game.

Examples of Zero-Sum Games

To better understand zero-sum games, let's consider some straightforward examples:

- **Poker**: In a game of poker, the amount of money won by some players is exactly equal to the amount lost by others. The total amount of money in the game (the "pie") does not change; it merely changes hands based on the outcome of the game.
- **Chess and Sports Competitions**: Games like chess or competitive sports such as soccer or basketball are also zero-sum games. One team's victory is another team's loss. The result is a ranking or score that transfers status or advancement from the losers to the winners.
- **Trading on a Stock Exchange**: Day trading in the stock market can also be seen as a zero-sum game, especially in the short term. If one trader profits from buying low and selling high, another trader will incur a loss. The total monetary value in the system remains constant; it is the distribution among the traders that changes.

Theoretical Framework of Zero-Sum Games

In the theoretical framework of zero-sum games, mathematical models are used to predict outcomes based on the strategies employed by different players. These models assume rational players who are fully aware of the game's rules and the payoffs and losses associated with each outcome.

Game Theory and Strategy Development

In zero-sum games, strategic development involves anticipating the moves of your opponents and countering them effectively. Since each player's goal is to maximize their own payoff at the expense of others, understanding and predicting the strategies of your opponents is crucial.

Minimax Strategy: A common approach in zero-sum games is the minimax strategy, where a player minimizes the potential maximum loss. This strategy is particularly used in two-player games and involves choosing a strategy that minimizes the worst possible outcome, thus protecting against the maximum potential loss.

Implications of Zero-Sum Thinking

Understanding zero-sum games is not just about recognizing when you are in a competitive scenario; it's also about understanding the limitations of zero-sum thinking. In real life, not all situations are zero-sum, and approaching every scenario as if it were can lead to missed opportunities for collaboration and value creation.

Negotiations and Business Deals: While some negotiations might seem like zero-sum games, often, they are not. Approaching business deals with a zero-sum mindset can lead to overly aggressive tactics that might spoil a potentially beneficial relationship or opportunity for collaboration.

Defining Non-Zero-Sum Games: Potential for All Players to Benefit or Lose

In non-zero-sum games, the outcomes are not strictly opposing, and the total value generated can vary, allowing possibilities for all players to benefit or for all to lose. This type of game reflects a large portion of real-life interactions, from business collaborations to diplomatic negotiations, where mutual gains are often possible and desired.

Understanding Non-Zero-Sum Games

Non-zero-sum games are scenarios where the sum of gains and losses among players does not necessarily equal zero. These games represent a more complex and realistic form of interaction where not every gain or loss is mirrored by an opposite loss or gain in others.

Characteristics of Non-Zero-Sum Games

Variable Outcomes: Unlike zero-sum games where the pie is fixed, non-zero-sum games can have outcomes where the total 'pie' increases or decreases, reflecting situations in the real world where cooperation can lead to greater overall benefits or where collective failures lead to shared losses.

Potential for Mutual Benefit: Players in non-zero-sum games can often find strategies that benefit all involved, unlike in zero-sum games where one player's gain is necessarily another's loss.

Interdependence: These games highlight the interdependence of players' strategies and outcomes. The decisions of one player usually affect the payoffs for others, not just in terms of gains being taken from others but in potentially enhancing the overall gains for all.

Examples of Non-Zero-Sum Games

To illustrate non-zero-sum games, let's explore a variety of examples across different fields:

Business Collaborations: When companies collaborate on a project, such as developing new technology, they create a non-zero-sum game where both can benefit from the success of the project. The success of the collaboration can lead to increased market size, shared profits, and enhanced reputations, increasing the 'pie' for both parties involved.

Environmental Agreements: International agreements on climate change are also non-zero-sum. When countries cooperate to reduce global emissions, the benefits include reduced global warming, healthier populations, and stabilized ecosystems, which are shared globally, even though the cost and effort of reducing emissions are distributed unequally.

Negotiations: Many negotiation scenarios are non-zero-sum because negotiators can often create value that benefits all parties. For instance, in a job salary negotiation, an employer and a potential employee might negotiate a package that includes flexible working hours and professional development opportunities, which are not strictly monetary but create high value for the employee, while the employer retains a valuable employee without excessively increasing salary expenses.

Theoretical Framework of Non-Zero-Sum Games

In non-zero-sum games, the theoretical framework often involves cooperative game theory, which studies how cooperation can improve the outcomes for all players compared to non-cooperative scenarios.

Cooperative Game Theory

This branch of game theory looks at how binding agreements and coalitions between players can lead to collectively better outcomes. It uses concepts such as the core, which is the set of possible distributions of total gains that cannot be improved upon by any subset of players forming a coalition.

Example: In a cooperative business venture, analyzing the game might involve looking at how different contributions by each company (such as technology, brand name, or distribution networks) contribute to the overall success and how the profits should be fairly divided to keep all parties satisfied and cooperative.

Strategic Implications in Non-Zero-Sum Games

Understanding non-zero-sum games is crucial for developing strategies that seek mutual benefit, which is often key in modern economic and international relations. Recognizing that cooperation can lead to better outcomes for all parties changes the approach from competitive to collaborative strategies.

- **Long-term Relationships and Trust**: Building long-term relationships and trust becomes critical in non-zero-sum games, as ongoing cooperation can lead to continued mutual benefits. This is evident in industries where repeat business and long-term contracts are pivotal, and where the goodwill generated by fair play can lead to more business opportunities.
- **Innovation and Creative Solutions**: Non-zero-sum games encourage innovative and creative solutions that expand the pie rather than just redividing it. This is crucial in fields like technology and pharmaceuticals, where collaboration often leads to breakthrough innovations that create new markets or dramatically expand existing ones.

Implications of Game Type: How the Nature of the Game Affects Strategy

In game theory, understanding whether you are engaged in a zero-sum or a non-zero-sum game profoundly affects your tactical approach. This understanding extends beyond mere theoretical interest and has practical applications in various areas of life, including economics, business, international relations, and personal decision-making. By comprehensively examining how the characteristics of these games influence strategic behavior, you will be equipped to more effectively navigate and influence outcomes in diverse scenarios.

Understanding the Strategic Implications of Zero-Sum Games

In zero-sum games, where one player's gain is another's loss, strategies tend to be competitive rather than cooperative. The rigid structure of these games where the total value remains constant implies that the primary objective is to maximize one's own payoff at the expense of others.

Competitive Dynamics

In zero-sum environments, players often adopt aggressive and defensive strategies aimed at maximizing their advantage. For example, in a bidding war on a contract, companies will push their limits to offer the most attractive bid while ensuring that their rivals do not secure the contract, often resulting in a tight margin.

Risk Management

Managing risk becomes crucial as any loss directly benefits competitors. Players must carefully assess the risk-reward ratio of their strategies, as a misstep can lead to a direct advantage for their opponents. In investment trading, for instance, traders must not only aim to maximize returns but also be vigilant of market movements that could advantage their competitors.

Secrecy and Surprise

In zero-sum games, maintaining secrecy about one's strategies can provide a critical edge. Military tactics often rely on the element of surprise, which hinges on keeping operational plans confidential. Similarly, companies may keep their product development processes secret to prevent competitors from copying or sabotaging their market launch.

Exploring the Strategic Landscape of Non-Zero-Sum Games

Non-zero-sum games, where mutual gains are possible, encourage a more collaborative and integrative approach. These games recognize that all parties can benefit from cooperating, or conversely, all can suffer from competition.

Collaborative Strategies

The potential for increasing the total pie or creating mutual benefits leads to strategies that emphasize cooperation, negotiation, and fair trade. In business, for instance, joint ventures between companies can lead to shared technological advancements, opening up new markets that were not accessible to individual firms alone.

Trust and Relationship Building

Long-term relationships and trust become valuable assets in non-zero-sum games. Strategies often involve building alliances and partnerships that can provide sustained benefits over time. Diplomatic relations between countries, for example, focus on building alliances that can lead to economic, security, and cultural exchanges benefiting all involved parties.

Flexibility and Adaptation

The dynamic nature of non-zero-sum games requires strategies that are adaptable to changing circumstances and opportunities. Players must be ready to adjust their approaches based on ongoing feedback and the evolving interests and capabilities of all participants.

Strategic Implications in Mixed-Game Environments

In real-world scenarios, players often encounter mixed-game environments where elements of both zero-sum and non-zero-sum games are present. This complexity requires a hybrid approach to strategy development.

Balancing Competition and Cooperation

Understanding when to compete and when to cooperate is crucial in mixed-game scenarios. For example, two companies may compete fiercely in one market segment while collaborating in another area where their interests align, such as environmental sustainability efforts.

Navigational Acuity

Players must develop an acute sense of judgment to navigate mixed games effectively. This involves continuously analyzing the game dynamics to understand when shifts occur between zero-sum and non-zero-sum scenarios and adjusting strategies accordingly.

Ethical Considerations

Strategic decisions in mixed games also involve ethical considerations, as the impact of these decisions can extend beyond the immediate players to affect broader communities and environments. Leaders must consider the ethical implications of their strategic choices, balancing the pursuit of competitive advantage with responsibility toward stakeholders and society.

Real-World Examples: Situations Illustrating Both Types of Games Across Various Fields

Zero-Sum and Non-Zero-Sum Games in Economics

Zero-Sum Example: Commodity Trading In the world of commodity trading, one trader's gain is often directly linked to another's loss. For instance, if one trader bets on the price of oil rising by purchasing futures contracts, and another bets on the price falling, any price movement results in a win for one and a loss for the other based on the contract's zero-sum nature.

Non-Zero-Sum Example - Economic Growth through Innovation When a company innovates a new product that improves efficiency or enhances quality of life, it does not just steal market share from competitors but can expand the market itself. For instance, the introduction of smartphones expanded the overall market for mobile communication and computing, creating benefits for consumers and numerous businesses beyond just the initial innovators.

Zero-Sum and Non-Zero-Sum Games in Politics

Zero-Sum Example: Electoral Politics In electoral politics, particularly in systems where there is a single winner, such as a presidential election, the gain of one candidate in terms of votes is a direct loss to another. Each vote for one candidate reduces the potential votes for another, highlighting the zero-sum nature of electoral contests.

Non-Zero-Sum Example: Legislative Compromise When political parties negotiate a legislative compromise, the outcome can be non-zero-sum. By reaching an agreement on a contentious issue, all parties can claim a win by showing effectiveness to their constituents, even though the final legislation may require concessions from each side.

Zero-Sum and Non-Zero-Sum Games in Personal Life

Zero-Sum Example: Divorce Settlements In a contentious divorce where partners are vying for a limited amount of assets, the division often becomes a zero-sum game. Each asset awarded to one partner means fewer assets available to the other, making the process inherently adversarial.

Non-Zero-Sum Example: Parenting Decisions When parents decide on the best approach to educating their child, they can create a non-zero-sum scenario. By discussing and integrating their views, they can develop a parenting strategy that combines their strengths and meets the child's needs better than either parent could alone.

Zero-Sum and Non-Zero-Sum Games in Business

Zero-Sum Example, Price Wars In competitive markets, price wars represent a zero-sum game where businesses compete for market share by undercutting each other's prices. Each dollar decrease in price meant to attract more customers directly reduces profit margins, putting businesses against each other in a direct financial tug-of-war.

Non-Zero-Sum Example: Strategic Alliances When companies form strategic alliances, they collaborate to achieve goals that would be difficult to achieve independently. For instance, a tech company and an automotive company might partner to develop autonomous vehicles, pooling resources to innovate more effectively than either could on its own.

Zero-Sum and Non-Zero-Sum Games in Sports

Zero-Sum Example: Tournament Outcomes In a knockout tournament, the advancement of one team to the next round directly results from the elimination of another team. Each match's outcome directly transfers the opportunity to advance from the loser to the winner, encapsulating the zero-sum nature.

Non-Zero-Sum Example: Sportsmanship and Game Development While individual matches might be zero-sum, the broader context of sports like promoting sportsmanship and developing the game can be non-zero-sum. Programs that aim to teach youth the values of teamwork and fair play benefit all participants, as well as the sport itself, by fostering a positive culture and increasing participation.

Zero-Sum and Non-Zero-Sum Games in Technology

Zero-Sum Example: Patent Races In technology development, companies often race to patent innovative technologies. Obtaining a patent can be zero-sum because securing a patent on a new technology can prevent competitors from using that technology, making the gain of one company the loss of opportunity for others.

Non-Zero-Sum Example: Open Source Software Development The development of open-source software is inherently non-zero-sum because contributions improve the software's quality and capabilities, which benefits all users and developers, regardless of their contributions. This collaborative approach contrasts sharply with the competitive dynamics of patent races.

Zero-Sum and Non-Zero-Sum Games in Social, Environmental, and Legal Contexts

Zero-Sum Example: Litigation In litigation, particularly in civil cases, one party's gain (in terms of compensation or a favorable verdict) generally comes at the expense of the other party. The adversarial nature of most legal disputes underscores their zero-sum nature.

Non-Zero-Sum Example: Community-Led Environmental Conservation Environmental conservation efforts led by communities often result in non-zero-sum outcomes where actions to preserve a local ecosystem benefit all community members. Initiatives like community forests or local wildlife conservation are collaborative efforts where everyone gains from a healthier environment.

Reflection Questions

1) Understanding Game Types: Reflect on a situation from your experience where you were involved in a game (this could be a negotiation, a business deal, or even a family decision). Was it a zero-sum or a non-zero-sum game? How did the nature of the game affect your strategy and the outcome?

2) Strategic Differences: How would your approach to the above situation change if the game's nature were reversed (from zero-sum to non-zero-sum or vice versa)? What strategic adjustments would you consider to maximize your benefits or minimize losses?

3) Real-World Analysis: Choose one of the real-world examples provided in this section that resonates with you. Analyze the implications of the game being zero-sum or non-zero-sum. What lessons can be drawn about how individuals or organizations should tailor their strategies based on the type of game they are engaged in?

Part 3 - Advanced Strategies and Applications

While we proceed in the exploration of game theory, it becomes clear that games can be categorized across different axes. Some of these axes we have already discussed as part of core game theory concepts, and others will be explored in the following sections. Below, you will find a diagram that exemplifies the key axes.

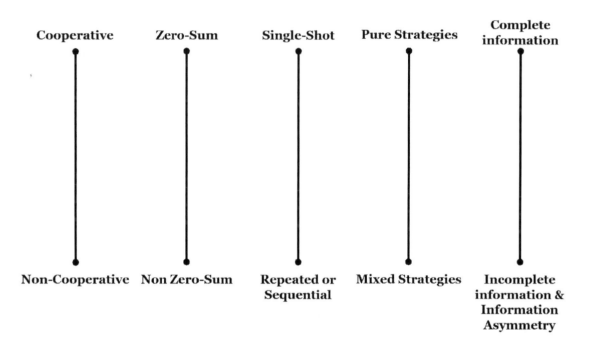

Section 7: Repeated and Sequential Games

Definition and Significance of Repeat Interactions

Unlike single-shot games where players interact once and the game ends, repeated games involve players interacting multiple times, offering a rich framework for analyzing how long-term relationships and strategies evolve. This section will go into the definition, mechanics, and significance of repeated games, providing insights into how they shape behaviors and decisions in various real-world contexts.

Defining Repeated Games

Repeated games, or iterated games, occur when the same game is played multiple times by the same participants. Each round of the game provides players with information that can influence their strategies in subsequent rounds. The key characteristic of repeated games is the continuity of interaction, which introduces a dynamic of past actions influencing future actions.

Core Elements of Repeated Games
- **The Stage Game**: This is the basic game that is played repeatedly. It can be any game, such as the Prisoner's Dilemma, where the players' choices and the outcomes in each iteration are defined.

- **Temporal Component**: Repeated games are defined by their ongoing nature. The number of repetitions can be finite (known in advance) or infinite (unknown or indefinite).
- **Information Flow**: Depending on the game's setup, players may remember previous outcomes and base their current decisions on past interactions, leading to a history-dependent strategy.

Significance of Repeat Interactions

The repetition of interactions in these games introduces several strategic considerations that are absent in single-shot games. Here are some of the reasons why repeated games are significant in both theoretical and practical contexts:

Building Trust and Cooperation

In repeated games, the possibility of future retaliation or reward can encourage cooperation even in scenarios like the Prisoner's Dilemma, where defection would be the dominant strategy in a single-shot version. Over time, players learn about each other's behaviors and can develop strategies based on trust and reciprocity.

Strategy Development Over Time

Players in repeated games can adjust their strategies based on the outcomes of previous rounds. This dynamic allows for more sophisticated strategies, such as "tit-for-tat" in the Prisoner's Dilemma, where a player initially cooperates and then replicates the opponent's previous move in subsequent rounds.

Punishment and Forgiveness

Repeated interactions allow strategies involving punishment for defection and forgiveness following cooperation. These strategies can enforce stable cooperation patterns and discourage deviations that might be profitable in a single-shot context.

Long-Term Payoffs

In repeated games, the focus often shifts from short-term gains to long-term payoffs. Players might endure short-term losses to establish a reputation or relationship that yields greater cumulative benefits.

Theoretical Framework of Repeated Games

Understanding repeated games involves analyzing how strategies evolve over time and how they depend on the game's temporal structure and the players' memory.

Finite vs. Infinite Games

The analysis often differentiates between games repeated a finite number of times and those with an indefinite or infinite horizon. The endgame strategies can differ significantly, with infinite games potentially fostering more cooperation due to the uncertainty about when interactions will end.

Perfect vs. Imperfect Information

In some repeated games, players have perfect information about past actions, while in others, some or all past actions are unknown. The level of information availability can dramatically affect the strategies, with more information generally leading to more accurately tailored strategies.

Sequential Games

Sequential games are characterized by the order in which players make their moves. Unlike simultaneous games, where all players make their decisions at the same time without knowledge of others' choices, sequential games allow players to observe the actions of those who preceded them before making their own decisions. This structure introduces a layer of strategic depth and foresight into the game.

Core Elements of Sequential Games

- **Order of Play**: In sequential games, the order in which players take their turns is fixed and known to all players. Each player's decision can depend on the observable actions of the players who have already moved.
- **Decision Nodes**: The game can be represented as a tree of decisions, where each node represents a point at which a player must make a choice. The path from the root to any terminal node (an endpoint of the tree) represents a complete sequence of moves.
- **Information Sets**: These are critical in understanding what players know at the time they make decisions. An information set for a player includes all the decision points they could face with the same knowledge about preceding moves.

Significance of Sequential Games

The structure of sequential games offers insights into dynamic decision-making processes where the timing and sequence of actions play crucial roles. Here's why understanding these games is significant:

Strategic Depth and Forward Thinking

Players in sequential games must anticipate the future moves of their opponents. This requires not only reacting to the current state of the game but also strategic planning and forward-thinking, considering how today's decisions will impact future possibilities.

Influence of Commitment and Credibility

In sequential games, earlier players can commit to certain strategies that influence the choices of later players. The credibility of these commitments can significantly affect the strategies of subsequent players, as they base their decisions on the expected actions of those who have already moved.

Importance of Signaling

Players can use their moves to send signals to other players about their intentions, strength, or type of strategy they are employing. Understanding how to interpret these signals is crucial for later players when it's their turn to act.

Theoretical Framework of Sequential Games

To effectively analyze sequential games, game theorists use a variety of conceptual tools that help illuminate the strategic structure of these interactions.

Extensive Form Representation

Sequential games are often represented in extensive form, which is a tree diagram showing the sequence of moves, the choices at each decision point, the players who make those decisions, and the payoffs received at the end of the game for every possible combination of actions.

Backward Induction

A common method used to solve sequential games is backward induction, where one starts from the end of the game and works backwards to determine the optimal strategy at each preceding decision point. This process reveals the subgame perfect equilibrium, where the strategy represents an optimal decision at every point in the game.

Subgame Perfection

This concept addresses the credibility of strategies in sequential games. A subgame perfect equilibrium is a refinement of Nash Equilibrium which requires that the players' strategies constitute a Nash Equilibrium not only in the whole game but also in every subgame. This eliminates non-credible threats and promises.

Differences from Single-Round Games: Impact on Strategy and Decision-Making

Sequential games differ fundamentally from single-round games in terms of strategy formulation and decision-making processes. Understanding these differences is crucial to navigate complex strategic environments effectively, whether in business, personal interactions, sports, or politics. The primary distinction between single-round games and their repeated or sequential counterparts lies in the depth and complexity of the strategic considerations involved. Here's how these differences fundamentally alter the game's nature:

1. Memory and History

In single-round games, each game stands alone with no history or future. Decisions are made based solely on the information and strategies relevant at that moment. In contrast, repeated and sequential games incorporate the history of past interactions, which can significantly influence future strategy and decisions. This historical awareness allows players to adjust their strategies based on past outcomes, fostering a dynamic interaction where strategy evolves over time.

2. Forward-Looking Behavior

Single-round games require no consideration of future consequences beyond the immediate game. However, in repeated and sequential games, players must consider the future implications of their current actions. This forward-looking behavior encourages more strategic depth, as decisions now must account for future rounds and the potential responses from other players.

3. Retaliation and Reward

The ability to retaliate against or reward other players in future interactions is absent in single-round games. In repeated games, however, the possibility of retaliation for defection or rewards for cooperation exists, altering how strategies are developed. Players may refrain from exploiting others even in tempting situations to maintain a cooperative relationship or avoid future retaliation.

4. Building Reputation and Trust

In repeated interactions, players have the opportunity to build reputations that can significantly affect their strategic possibilities. A reputation for being trustworthy or tough on defection can lead to more favorable outcomes in the long run. In single-round games, reputation plays no role, as there is no continuation of relationships beyond the immediate game.

5. Strategy Complexity and Sophistication

Strategies in single-round games are generally straightforward, as they do not need to account for any subsequent interactions. In repeated and sequential games, strategies become more complex and sophisticated, involving contingencies that depend on the sequence of earlier moves and potential future turns. This complexity allows for richer strategic interactions and more nuanced decision-making.

Real-World Examples: Situations of Repeated and Sequential Games Across Various Fields

Repeated and Sequential Games in Economics

Economic Policy Negotiations

Repeated games are central in economic policy negotiations where countries repeatedly interact over issues like trade agreements or monetary policies. Each round of negotiation builds on the outcome of the previous, allowing countries to adjust their strategies based on past interactions, fostering a dynamic where long-term strategy, rather than immediate gains, guides decisions.

Market Competition

In market competition, businesses frequently engage in sequential games. Consider the release cycles of technology companies that are staggered and respond to competitor moves. Each product launch is a move in a sequential game where businesses anticipate competitors' innovations and time their releases to maximize market impact and capture.

Repeated and Sequential Games in Politics

Politicians must consider how their actions will be perceived in the context of ongoing political relationships and future electoral considerations, diverging significantly from single-issue negotiations.

Legislative Processes

The process of passing legislation often involves sequential games, where the order of moves—such as proposing amendments and voting—can significantly influence the outcome. Politicians must strategize not only on the content of the legislation but also on the timing of various actions to align with or disrupt the agendas of opponents.

Election Campaigns

Candidates adjust their strategies in response to the actions and policies of their opponents, trying to position themselves advantageously in the eyes of the electorate as the campaign unfolds.

Election strategies are repeated games where past election results influence future strategies. Political parties adjust their policies and campaign tactics based on the outcomes of previous elections, learning which strategies maximize voter support and adapting over time.

Repeated and Sequential Games in Personal Life

In personal relationships, repeated interactions are the norm, and understanding how to manage these dynamics can lead to more harmonious relationships. Strategies that consider the long-term health of the relationship, such as compromise and reciprocity, become crucial.

Family Dynamics

Family relationships often involve repeated games where actions and outcomes in earlier interactions affect future behaviors. Trust, a crucial element in these relationships, develops from repeated positive interactions, while a single betrayal can shift the dynamics significantly, influencing future interactions.

Negotiations in Everyday Life

Negotiations over household responsibilities or personal finances between couples or roommates are examples of sequential games where the order of concessions and compromises can affect the overall satisfaction and outcomes of the negotiations.

Repeated and Sequential Games in Careers

Professional Development and Promotions

Career advancement is a sequential game where individuals take steps over time that build on each other. Decisions about job changes, additional training, or taking on new projects are made with an eye on future career opportunities and potential advancements.

Workplace Negotiations

Negotiations over salary or project assignments are often repeated games where past negotiations set precedents and influence future discussions. Employees and employers must consider the long-term implications of each negotiation, not just the immediate outcomes.

Repeated and Sequential Games in Business

Companies must consider how actions taken today affect their long-term reputation and relationships with stakeholders, contrasting sharply with single-transaction strategies that prioritize immediate gains.

Strategic Business Alliances

Business alliances, such as partnerships in research and development, are repeated games where each collaboration builds trust and sets the stage for future joint ventures. Each project's outcome influences the likelihood and nature of future cooperation.

Pricing Strategies

Sequential games are evident in pricing strategies where businesses adjust their pricing based on competitor actions and market responses, carefully timing price drops or increases to maximize profitability.

Business Negotiations

Each party observes the other's offers and adjusts their strategies accordingly, aiming to optimize their outcomes based on the evolving dynamics of the negotiation.

Repeated and Sequential Games in Sports

Seasonal Sports Strategies

In sports leagues, the strategy for a season is a repeated game where the outcomes of early matches influence the strategies for later games. Teams adjust their tactics based on their standing and the performance of others as the season progresses.

Coaching Decisions

Coaching strategies, such as player substitutions and play calls, are often sequential, with each decision building on the game's current state and anticipating future plays and opposing strategies.

Repeated and Sequential Games in Technology

Innovation Cycles

Technology development is a sequential game where companies time their innovations to stay ahead of competitors. Each new product release or update builds on previous versions and market feedback, influencing future development cycles.

Platform Wars

Competitions between technology platforms, like mobile operating systems or video streaming services, are repeated games where each iteration or update can change competitive dynamics, influencing user choices and future market shares.

Repeated and Sequential Games in Social Dynamics
Social Influence and Conformity
Social behaviors, such as trends in fashion or public opinion on social issues, often unfold as sequential games where early adopters influence later behaviors and societal norms evolve over time through repeated interactions.

Group Projects
Collaborations on group projects, whether in academic or professional settings, involve repeated games where the contribution of each member in early phases affects the dynamics and contributions in later phases, influencing the overall success of the project.

Repeated and Sequential Games in Environmental Issues
Resource Management
Environmental conservation efforts, like managing fisheries or forests, are repeated games where the sustainability of the resource depends on the actions taken in each period, with past exploitation levels and regeneration rates influencing future decisions.

Climate Change Negotiations
International negotiations on climate change are sequential games where each summit builds on the agreements and commitments of previous meetings, with nations adjusting their strategies based on the actions and commitments of others.

Repeated and Sequential Games in Legal Contexts
In legal settings, particularly in litigation or long-term contractual relationships, the understanding of sequential and repeated games can inform strategies that consider the broader implications of legal decisions and interactions, influencing approaches to negotiation and conflict resolution.

Litigation Strategies
In legal disputes, particularly those involving multiple hearings or trials, strategies are often sequential, with each side adjusting their approach based on the outcomes of earlier motions or rulings. In addition, the order of witness testimonies and evidence presentation can significantly affect the jury's perception and the trial's outcome, illustrating the strategic sequencing in legal settings.

Contract Negotiations
Negotiations over contracts are repeated games where each party's willingness to compromise in current negotiations can influence the terms and cooperation levels in future contractual relationships.

Reflection Questions
1) Evaluating Repeated Interactions: Think of a situation in your personal or professional life where you repeatedly interact with the same individuals or groups. How do these repeated interactions affect your

strategic decisions? Reflect on whether a long-term cooperative strategy or a competitive strategy has been more effective, and why.

2) Contrasting Game Types: How would your strategy change if a repeated game you're currently involved in (like a project team at work or a weekly poker game) were to be played only once? What factors would you consider in a single-round game that you might overlook in a repeated game?

3) Real-Life Sequential Games: Choose a real-world example of a sequential game from this section. Describe how the order of moves impacts the strategies and potential outcomes. How does this sequential structure affect the decisions of each player differently than if all moves were made simultaneously?

Section 8: Mixed Strategies and Randomness

What are Mixed Strategies?

Understanding Mixed Strategies
In game theory, mixed strategies involve choosing between different actions randomly instead of picking one single action every time (aka pure strategies). This means a player doesn't always stick to one predictable strategy but instead switches among several strategies based on certain probabilities.

Key Elements of Mixed Strategies
Probability Distribution: Each action or strategy a player can choose is assigned a specific probability that tells how often it should be selected. The total of all probabilities must add up to 100%, ensuring a strategy is chosen each time.

Randomization Device: To implement a mixed strategy fairly, players often use a device or method that helps them make random choices, like rolling a dice or using a computer program, or just changing their choice autonomously.

Indifference Principle: The goal of using mixed strategies is to make an opponent unsure about how to respond. Ideally, no matter what the opponent thinks you will do, they can't find a way to consistently beat you, making them indifferent to their own choice of strategy.

Why Mixed Strategies Matter
Mixed strategies reflect the complex decision-making needed in real-life situations where outcomes depend on interactions between different players' choices and some element of chance.

By mixing up strategies, a player introduces uncertainty into the game, making it hard for opponents to predict and counter their moves effectively. This can help protect the player from being exploited by a savvy opponent.

Using mixed strategies allows players to adapt more fluidly to changing situations in a game. They can shift their approach as needed, which is vital in dynamic environments where sticking rigidly to one strategy might lead to failure.

In games where no single strategy guarantees a win every time (games without a clear winning strategy for all scenarios), mixed strategies help players find a balance that maximizes their potential gains or minimizes losses over time.

Theoretical Insights into Mixed Strategies

To get a grip on mixed strategies, it's useful to look at how they're formulated in theory:

Nash Equilibrium with Mixed Strategies

Even in games where no single strategy is best, John Nash's theory suggests there's a set of mixed strategies that can stabilize the game. Each player chooses their strategy mix in a way that no one can better their position by unilaterally changing their own mix.

Calculating Expected Payoffs

Analyzing mixed strategies involves calculating what each player can expect to win or lose on average, considering the mix of strategies they and their opponents are using. This calculation helps players decide which mix might work best for them.

Using Minimax in Competitive Games

In highly competitive or zero-sum games, where one player's gain is another's loss, using mixed strategies can help a player minimize their potential maximum loss. This approach is about making safe choices that ensure the best outcome in the worst-case scenario.

Role of Randomness in Games: How Unpredictability Affects Game Outcomes

Understanding Randomness in Games

Randomness in games refers to elements or events that are out of the players' control and whose outcomes are unpredictable. These elements can range from dice rolls and card draws to algorithmically generated random events in digital games. The inclusion of randomness means that even with a perfect strategy, the outcome can still be uncertain due to these chance elements.

Key Characteristics of Randomness in Games

Uncertainty: Randomness injects a degree of uncertainty into games, which can complicate players' ability to predict outcomes and plan strategies.

Variability: It introduces variability in outcomes, ensuring that no two games are exactly alike, even with the same strategies and players.

Excitement and Challenge: Randomness can make games more exciting and unpredictable, enhancing the challenge by requiring players to adapt to continuously changing conditions.

Impact of Randomness on Strategy

The inclusion of randomness in games affects how strategies are formulated and executed. Players must consider not only the actions of other players but also the possible outcomes of random events.

Adaptation and Flexibility

Players must be adaptable and flexible in their strategies to accommodate the potential range of outcomes that randomness might produce. This often involves preparing for the worst-case scenarios while hoping for the best-case outcomes.

Risk Management

Effective strategy in random games often involves significant risk management. Players must assess the probability of various outcomes and decide whether the potential rewards justify the risks. This balance is crucial in games where high-stakes decisions are influenced by chance.

Probability and Expected Values

Strategies often rely on probability calculations to make informed decisions. Players evaluate the expected values of different actions, which are calculated by multiplying the outcomes by their probabilities and summing the results. Strategies that maximize expected value tend to be more successful over time.

The Role of Randomness in Different Types of Games

Randomness plays varying roles depending on the type of game, from classic board games to complex economic simulations.

Board Games

In board games like Monopoly or Risk, dice rolls are critical, affecting movement and outcomes of player interactions. The randomness from dice introduces an element of luck that can mitigate the advantages of strong strategic play.

Card Games

Card games like poker or blackjack involve randomness in the dealing of cards. Players must constantly adapt their strategies based on the hand they are dealt and the potential cards to come, making probability a key component of their strategic approach.

Sports

Many sports incorporate randomness, such as environmental conditions affecting the game or random referee decisions. Players and teams must adapt their tactics not only to the opposing team's strategy but also to these unpredictable elements.

Economic and Simulation Games

In economic simulation games, randomness can simulate market fluctuations or economic crises, requiring players to adjust their strategies in response to unexpected changes in the game environment.

Developing Mixed Strategies: Empowering Your Strategic Planning with Randomness

Embracing the Power of Randomness in Strategy

In the dynamic landscapes of business, politics, sports, and personal life, unpredictability is a constant. Mixed strategies provide a robust framework that leverages this unpredictability to your advantage, turning potential challenges into opportunities for strategic innovation.

Mixed strategies involve diversifying your tactical approaches by incorporating randomness into your decision-making processes. This method goes beyond conventional strategies by allowing for flexibility and adaptability, ensuring that your moves are less predictable and more fluid.

Integrating randomness means that instead of always responding in a predictable manner, you incorporate a range of possible actions at varying probabilities. This approach does not rely solely on chance; it's a calculated method of keeping your adversaries continually guessing and adapting.

Crafting Your Randomization Blueprint

Developing an effective mixed strategy involves careful planning and a deep understanding of the probabilities associated with different outcomes. It's about balancing risk and reward in a way that aligns with your overall goals.

Creating a Strategic Probability Matrix

Begin by mapping out all possible strategies and the outcomes they could potentially lead to. Assign probabilities to each strategy based on how often you think they should be executed to maximize your strategic advantage. This matrix will serve as your blueprint for implementing mixed strategies.

Utilizing Tools for Randomness

Adopt tools that can facilitate randomness in your strategy execution—whether digital random number generators, algorithms, or even simpler methods like dice rolls or shuffled cards, depending on the context of your strategic game.

The Strategic Benefits of Mixed Strategies

Incorporating randomness into your strategic planning has multiple benefits, making your decisions more robust against competitive tactics.

Enhancing Flexibility and Responsiveness

With mixed strategies, you can quickly adapt to the shifting dynamics of your environment. This responsiveness is crucial in high-stakes settings where flexibility can mean the difference between success and failure.

Preventing Predictability

By randomizing your actions, you effectively shield your strategic intentions from opponents, preventing them from developing counter-strategies that could undermine your objectives.

Optimizing Long-Term Outcomes

Mixed strategies allow for a more dynamic approach to achieving your goals, optimizing long-term outcomes by continuously adapting to new information and conditions as they arise.

Empowering Your Decision-Making

As you integrate mixed strategies into your strategic planning, you empower yourself to make decisions that are not only proactive but also highly adaptive to the complexities of the modern world. This empowerment comes from knowing that you can navigate uncertainty with confidence, turning each challenge into a strategic opportunity.

Real-World Examples: Embracing Mixed Strategies and Randomness Across Various Domains

Whether you're navigating the unpredictable waters of economics, engaging in politics, managing personal decisions, or aiming for victory in sports, the intelligent integration of randomness and mixed strategies can empower you with unmatched flexibility and strategic depth. This exploration will illustrate how mixed strategies and randomness are applied in real-world scenarios, transforming potential challenges into strategic advantages.

Mixed Strategies in Economics

In the complex world of economics, uncertainty is a constant companion. Here, mixed strategies offer a way to address market unpredictabilities and competitive behaviors effectively.

Stock Market Trading

Traders often use mixed strategies when deciding on a portfolio of investments. Randomness in selecting stocks or timing purchases can prevent predictable patterns that might be exploited by other market participants. By randomizing the entry and exit points within the market, traders can avoid price manipulations and benefit from unexpected market movements, optimizing their returns.

Competitive Pricing Strategies

Businesses frequently face the challenge of setting prices in competitive environments. A mixed strategy might involve occasionally changing the price of a product in unpredictable ways to avoid price wars with competitors. This randomization can prevent competitors from anticipating pricing moves, thus maintaining a healthy profit margin and market positioning.

Mixed Strategies in Politics

Politics is an area where unpredictability can be a powerful tool, influencing campaigns, negotiations, and policy-making.

Electoral Campaign Tactics

Political campaigns often employ mixed strategies in their messaging and public appearances. By varying the focus of campaign messages and the locations of rallies, candidates can appeal to a broader audience and keep opponents guessing about their next moves. This strategic variability ensures that their public engagements are both impactful and unpredictable, enhancing the campaign's reach and influence.

International Diplomacy

In international relations, countries might use mixed strategies in negotiations to secure better deals. By being unpredictable in their concessions and demands, nations can prevent other countries from exploiting their strategies, ultimately leading to more favorable diplomatic outcomes.

Mixed Strategies in Personal and Everyday Life

Even in personal life, randomness and mixed strategies can enhance decision-making, adding layers of complexity to interpersonal interactions and personal choices.

Daily Decision-Making

Individuals might use mixed strategies in their daily routines to optimize time and manage risks. For instance, varying commuting routes to avoid traffic congestion based on probabilistic information of traffic patterns can save time and reduce travel unpredictability.

Social Interactions

In social settings, people often employ mixed strategies in their responses and behaviors to maintain a balance in relationships. By sometimes being unpredictable in responses to friends' requests or invitations, individuals can manage their social commitments without setting rigid expectations.

Mixed Strategies in Career Development

Career paths are increasingly non-linear, with many professionals using mixed strategies to navigate their development and opportunities.

Job Application Processes

Professionals might apply to a range of positions, some aligned closely with their skills and others less so, to increase the probability of job offers. This approach involves a mix of safe choices and high-risk, high-reward applications, broadening the opportunities for career advancement.

Skill Development

Individuals often diversify their learning and skill acquisition, investing time in both their core professional area and in new, unrelated skills. This strategy prepares them for a wider range of career opportunities, making their career paths less predictable but more adaptable to changing job markets.

Mixed Strategies in Business

Businesses leverage mixed strategies to navigate competition, manage resources, and innovate within their industries.

Product Development

Companies often randomize elements of their product development processes to innovate effectively. This might involve varying investment levels in different R&D projects or randomizing the selection of project ideas to develop, ensuring a diverse portfolio that can adapt to market changes.

Marketing Campaigns

In marketing, companies frequently change their promotional strategies, randomly alternating between different media platforms, promotional offers, and target demographics. This strategy prevents market saturation and keeps the marketing impact fresh and effective.

Mixed Strategies in Sports

In sports, coaches and players utilize mixed strategies to gain competitive advantages during games and seasons.

In-Game Tactics

Sports teams often change their play styles unpredictably during a game to confuse opponents. A football team, for instance, might vary between aggressive attacks and defensive plays to unsettle the opposing team's strategy, optimizing chances of victory.

Season Long Strategies

Teams might also employ mixed strategies throughout a season, varying player line-ups and game strategies not only based on opponent but also in a randomized fashion to prevent predictability in longer tournaments.

Mixed Strategies in Technology

In the fast-evolving tech industry, companies employ mixed strategies to handle competition, innovation, and market dynamics.

Cybersecurity Measures

Firms randomize security measures, such as the timing of system audits and the methods used in penetration testing, to make it harder for attackers to predict and breach their systems. This randomness is crucial for maintaining system integrity against external threats.

Algorithm Development

Tech companies develop algorithms that incorporate randomness to enhance user experience or optimize operations. For example, search engines might randomly alter the algorithms that determine the display of search results to test different configurations and optimize user engagement and satisfaction.

Mixed Strategies in Social Dynamics

Social dynamics often involve complex interactions where mixed strategies can play a significant role in managing relationships and community engagements.

Community Engagement

Leaders in communities might vary their approaches to organizing events and initiatives to maintain engagement and participation. By not being predictable, they keep community interest high and ensure diverse attendance and involvement.

Mixed Strategies in Environmental Management

In environmental strategies, randomness can help manage resources sustainably and address challenges in conservation efforts.

Wildlife Conservation

Conservation strategies might involve randomizing the areas of focus for certain activities like patrols or scientific surveys to prevent poaching or interference. This strategy can increase the effectiveness of conservation efforts by making the enforcement pattern less predictable and more difficult to evade.

Mixed Strategies in Legal Practices

In the legal arena, attorneys often employ mixed strategies to enhance their case outcomes. Lawyers might vary their focus in legal arguments or the order in which evidence is presented to keep the opposing counsel off-balance, optimizing the chances of a favorable verdict.

Reflection Questions

1) Understanding Mixed Strategies: Reflect on a situation where you or someone you know faced multiple viable options and had to choose a strategy. How might employing a mixed strategy (using a randomized approach to decision-making) have altered the outcome? What scenarios might benefit from such an approach?

2) Role of Randomness: Think of a decision-making process you are involved in regularly, such as in business, sports, or personal finance. How could introducing randomness into your strategy help to improve outcomes or mitigate risks? What might be the challenges or downsides of incorporating randomness?

3) Analyzing Real-World Applications: From the examples of mixed strategies and randomness provided in this section, choose one that particularly interests you. Analyze how randomness and mixed strategies are utilized and discuss the impact of these approaches on the game's outcome. How does this example change your perception of strategic planning?

Section 9: Bayesian Games and Information Asymmetry

Introduction to Bayesian Games

Unlike the more straightforward settings you might be familiar with, where every player knows exactly what the game entails, Bayesian games introduce a layer of uncertainty that more closely mirrors the complexities of real-world decision-making. This section is your gateway to understanding how Bayesian games operate, and how you, as a player, can navigate these waters with strategic acumen.

Understanding the Core of Bayesian Games

At the heart of Bayesian games lies the concept of incomplete information. Imagine stepping into a game where not all the cards are visible—where each player has private information that significantly influences their decisions. In such games, you don't have full knowledge about other players' preferences, payoffs, or even their types (a term used to describe a player's characteristics or information). Each player's type is their own secret, and this uncertainty changes how you must approach the game.

The Nature of Incomplete Information

In Bayesian games, the incomplete information about other players is modeled through probability distributions. These are not just any guesses; they are educated estimations based on available information, and they significantly shape your strategic thinking. You, as a player, must consider not only your strategies and outcomes but also how to infer others' hidden information and predict their moves.

Navigating Through Uncertainty

The challenge and thrill of Bayesian games come from this very essence of not knowing all aspects outright. You must use the given, often limited, information to make the best decisions. This process involves continuously updating your beliefs and strategies based on the actions and signals from other players. It's like being a detective in a game, where each move provides clues to what the hidden information might be.

Formulating Strategies in Bayesian Games

How do you strategize in a game where key pieces of information are hidden? The answer lies in Bayesian Nash Equilibrium, a concept that extends the idea of Nash Equilibrium to situations with incomplete information. In such equilibria, each player's strategy maximizes their expected utility, given their beliefs about the other players' types and strategies.

This equilibrium is crucial because it reflects a state where no player can do better by unilaterally changing their strategy, given their beliefs about others' types and the strategies those types would choose. Your goal in such a game is not just to respond to what you see but also to what you believe others might be concealing. This involves a delicate balance of actions and reactions, predictions and adjustments.

The Role of Beliefs and Expectations

In Bayesian games, your beliefs about other players' types play a critical role in shaping your strategies. These beliefs are based on your own information and the overall setup of the game, including any common prior knowledge that all players share. As the game unfolds, you update these beliefs in response to new information gained from observing others' actions, which in turn affects your strategic choices.

The dynamic updating of beliefs is what makes Bayesian games particularly interesting and complex. You are constantly learning, adapting, and predicting. Each piece of new information can lead you to revise your strategies, making the game a dynamic and evolving puzzle.

Practical Implications of Bayesian Games

Bayesian games are not just theoretical constructs; they have practical implications in various fields such as economics, political science, and negotiation tactics. They help explain how economic agents make decisions under uncertainty, how politicians craft policies knowing they don't have all the information, and how businesses strategize when future markets are unpredictable.

Consider a market where companies must decide on investment strategies without full knowledge of competitors' plans or a negotiation where each party has private information about their willingness to compromise. In these cases, understanding Bayesian games can provide significant strategic advantages.

Understanding Information Asymmetry - Navigating the Uneven Terrain of Knowledge

The Essence of Information Asymmetry

Imagine you're entering a marketplace, but unlike a typical market, not everyone knows the same details about the products available. Some traders have insider knowledge about the quality of the goods they're buying or selling, while others must make decisions based on limited or skewed information. This scenario is not just a feature of marketplaces but is a common aspect of many strategic interactions in business, politics, and personal life.

Information asymmetry occurs when one party in a transaction has more or better information compared to another. This imbalance can significantly affect the outcomes of transactions, negotiations, and competitions. In your journey through strategic interactions, understanding and managing information asymmetry is crucial for ensuring that you're not at a disadvantage.

The Impact of Hidden Information

The challenges of information asymmetry are most vividly illustrated in situations where hidden information plays a critical role. Consider a used car market, famously discussed in economist George Akerlof's "The Market for Lemons". Sellers have more information about the quality of the car than buyers. This imbalance can lead to adverse selection, where the fear of buying a 'lemon' (a defective car) drives good cars out of the market, and only the bad ones remain.

"The Market for Lemons" can be explained simply by focusing on how the quality of goods and the imbalance of information between buyers and sellers can affect a market.

Imagine you're looking to buy a used car. There are good quality cars ("peaches") and bad quality cars ("lemons"). As a buyer, you find it difficult to tell the difference between a peach and a lemon before you buy and use the car. However, the seller knows exactly what they are selling.

Since buyers cannot tell which cars are lemons and which are peaches, they are only willing to pay a price that reflects the average quality of cars in the market. This average price is lower than what a peach should command, but higher than what a lemon is worth.

Sellers of good cars (peaches) know that their cars are worth more than this average price and might decide not to sell their cars at all, leaving only lemons for sale in the market. As a result, the quality of cars available in the market decreases because good cars are withheld from the market. This situation can spiral, worsening the average quality of cars available, and might eventually lead to the market for used cars collapsing.

Akerlof's theory shows how markets where sellers have more information than buyers can lead to high-quality goods disappearing from the market, which is detrimental to both buyers and the overall health of the market.

Your Challenge in Such Markets

When you face similar situations, whether buying a car, negotiating a job offer, or making investment decisions, the key lies in your ability to discern the true quality of what's on offer. Your strategy must involve either acquiring more information or cleverly interpreting the available signals to make the most informed decision possible.

Strategies to Overcome Information Asymmetry

To navigate environments riddled with information asymmetry, you need robust strategies that can either reduce the knowledge gaps or help you make the best out of the situation.

Seeking Information

Your first strategy might involve direct efforts to reduce information asymmetry. This could include conducting independent research, hiring experts, or using technological tools to gather data that is not readily available. For instance, before investing in a company, you might dive deep into its financial reports, read analyst insights, or even use satellite imagery to gauge the company's inventory levels.

Using Signaling and Screening

Another effective strategy is to use signaling and screening mechanisms. If you're on the less-informed side of the transaction, you can encourage the other party to reveal their information. For example, employers use diplomas and certifications as a screening tool to gauge the qualifications of potential employees without having to assess every candidate in depth.

Conversely, if you possess more information (like a high level of competence or a quality product), you can use signaling to differentiate yourself. This might involve obtaining certifications, warranties, or engaging in behaviors that credibly indicate your higher quality, thus distinguishing you from less informed or lower-quality competitors.

Storytelling in Strategy

In dealing with information asymmetry, storytelling becomes a powerful tool. Crafting a narrative around your product, service, or personal brand can influence perceptions and bridge information gaps. For instance, a company might share stories of customer satisfaction or product reliability that serve to reassure potential buyers about the quality of an offering.

Strategies in Bayesian Games: Mastering Decision-Making with Incomplete Information

Mastering strategies in Bayesian games empowers you to make informed decisions in situations laden with uncertainty and hidden information. By effectively updating your beliefs, signaling your type, screening for information, and committing to credible actions, you can navigate through and influence complex strategic environments. The essence of success in Bayesian games lies in your ability to adapt and respond to new information, turning the unknown into a strategic asset.

Understanding the Foundation of Bayesian Strategies

Bayesian games, characterized by their incomplete information, require a unique set of strategic considerations. Each player in a Bayesian game has private information, known as their 'type', which influences their choices but is not fully known to others. The challenge and skill in these games lie in making decisions that are optimal based on probabilistic beliefs about other players' types and potential actions.

Formulating Beliefs and Bayesian Updates

The cornerstone of strategy in Bayesian games is the formulation and updating of beliefs. As the game progresses and more actions are observed, you continuously update your beliefs about other players' types using Bayesian inference. This process involves adjusting your initial beliefs (priors) in light of new evidence from the actions others take, refining your strategy in response to this updated information.

Expected Utility Maximization

In Bayesian games, strategies aim to maximize expected utility, which is calculated based on the probability distribution over possible types of the other players and their corresponding actions. Your strategy should consider not only the immediate payoffs from different actions but also the expected future benefits, weighed by the likelihood of various outcomes.

Key Strategies in Bayesian Games

Navigating through the complexities of Bayesian games involves employing sophisticated strategies that can adapt to the evolving understanding of the game's environment.

Signaling and Screening

Signaling involves taking actions that credibly convey information about your type to other players, influencing their beliefs and subsequent actions. If you possess a high-quality attribute or information, you might choose actions that signal this quality to others, shifting their strategies in your favor.

Screening, on the other hand, is used by players who lack information. It involves setting up scenarios or making decisions that induce other players to reveal their private information. For example, a company might offer a variety of contracts designed to reveal the risk preferences of potential employees based on which contract they choose.

Commitment and Credibility

In many Bayesian games, the ability to commit to a strategy credibly can alter the expectations and actions of other players. For instance, a firm might commit to a certain level of output to influence market prices or competitor behavior. The credibility of such a commitment depends on the observable and verifiable nature of the action, and sometimes on the reputation or past behavior of the player.

Contingent Strategies

These strategies are conditional plans of action that depend on the updates to your beliefs about other players' types. As new information becomes available, contingent strategies allow you to respond optimally based on the latest set of beliefs. This adaptability is crucial in maintaining strategic flexibility and can be pivotal in environments that are highly dynamic and uncertain.

Real-World Applications of Bayesian Games and Information Asymmetry

Economics: Market Dynamics and Pricing Strategies

In market economics, your understanding of information asymmetry and Bayesian strategies is essential in shaping behaviors and outcomes. Consider a scenario where companies introduce new products with varying, undisclosed quality levels—known to producers but not to you, the consumer. This situation, as introduced earlier is known as a market for "lemons," can lead to adverse selection where inferior products displace superior ones as consumers adjust their spending based on the average expected quality.

Strategic Response: Companies might implement signaling strategies like offering warranties to credibly communicate their products' high quality. Meanwhile, you as a consumer could develop screening mechanisms such as preferring products from reputable brands or relying on third-party reviews to refine your expectations about product quality.

Politics: Campaign Strategies and Voter Behavior

Bayesian games capture the nuanced interactions between political candidates and voters, where candidates typically possess more detailed information about their planned policies. As a voter, you are tasked with deducing these policies from public statements and historical actions, creating a dynamic exchange where beliefs are updated continually throughout the campaign.

Strategic Response: Politicians might use signaling to influence your perceptions, often by adopting public stances on prominent issues or aligning with well-known figures. You might use a Bayesian approach to update your beliefs and voting strategies based on the consistency and credibility of these signals.

Personal Life and Everyday Decisions

Daily decisions often involve Bayesian reasoning, especially when full information is lacking. Situations like choosing a new mechanic or selecting a health insurance plan are typical examples where your past experiences and reviews become pivotal.

Strategic Response: You might use past interactions as Bayesian updates to inform future choices, such as selecting doctors based on personal experiences or recommendations, which serve as signals of quality and reliability.

Career: Job Market and Professional Relationships

The professional landscape is rife with information asymmetry, notably in hiring processes where employers and potential employees possess differing information sets. Employers may not know the true work ethic or skill level of candidates, and job seekers might be unaware of the actual work culture or expectations.

Strategic Response: Employers may utilize interviews, reference checks, and trial periods as screening tools. As a candidate, you can signal your capabilities and fit through personalized cover letters, portfolios, and strategic interviewing tactics to influence the beliefs of potential employers.

Business: Competitive Strategy and Corporate Espionage

In business, making decisions often requires dealing with incomplete information about competitors' strategies. Corporate espionage and competitive intelligence gathering are extreme measures that companies take to lessen information asymmetry.

Strategic Response: Companies may adopt mixed strategies in product development and market entry tactics to remain unpredictable to competitors, while also investing in counter-espionage measures and cybersecurity to safeguard their strategic information.

Sports: Coaching Tactics and Performance Enhancements

Coaching in sports typically involves making strategic decisions based on incomplete information about an opponent's future tactics. Coaches and players must predict potential changes in strategy based on historical data or in-game observations.

Strategic Response: Teams might employ a variety of strategies, sometimes altering their play based on Bayesian updates received from observing the opponent's early game tactics, effectively adapting to the dynamic nature of competitive sports.

Technology: Algorithm Design and Data Privacy

In technology, Bayesian games appear in scenarios like algorithm design, where developers must predict user behavior, which is often only partially observable. Data privacy issues, where companies know more about data usage than consumers, also reflect information asymmetry.

Strategic Response: Developers might use Bayesian inference to enhance user interface designs based on user interaction data. For data privacy, companies could adopt transparency as a signaling mechanism to reassure you about the security of your data.

Social Dynamics: Trust Building and Social Networking

Social interactions frequently incorporate Bayesian games, especially in trust-building scenarios where initial interactions help update beliefs about another's trustworthiness. This becomes even more pronounced online, where interactions and their genuine nature are less observable.

Strategic Response: You may signal your trustworthiness through consistent behavior over time or through mutual connections in social networking scenarios, influencing others' willingness to cooperate or engage.

Environmental Strategies: Conservation Efforts and Resource Management

In environmental management, information asymmetry often exists between regulators and companies about the true environmental impact of business activities. Bayesian strategies help estimate the likelihood of compliance based on past behavior and signals of corporate responsibility.

Strategic Response: Environmental regulators might use permits and regular inspections as screening tools, while companies could employ green policies as signals to suggest compliance and lessen regulatory scrutiny.

Legal Contexts: Negotiation and Litigation

In legal negotiations, information asymmetry can significantly impact the outcomes of cases, with one party often having more or better information about the case facts. Bayesian games unfold as each side updates its strategies based on new disclosures and evidence presented.

Strategic Response: Attorneys may use discovery processes to decrease information asymmetry and employ signaling through legal precedents or the strength of their legal team to influence settlement negotiations.

Legal Negotiations

In legal contexts, attorneys may use Bayesian strategies to decide whether to settle or proceed to trial. By assessing the opponent's type and potential case strength based on observable actions like previous litigation behavior or negotiation offers, attorneys can better strategize to achieve favorable outcomes.

Reflection Questions

1) **Recognizing Information Asymmetry**: Reflect on a recent situation, either personal or professional, where you had less information than others involved. How did this information asymmetry affect your decision-making process? In hindsight, how might you have use the learnings in this section to better handle the situation?

2) **Strategic Application of Bayesian Theories:** Consider a scenario in your work or personal life where making decisions with incomplete information is common (such as investing, hiring, or even dating). What strategies could you employ from Bayesian game theory to improve your outcomes? How does anticipating others' strategies based on their possible information sets change your approach?

3) **Real-World Analysis**: From the real-world applications discussed in this section, choose one example that resonates with you. How does the concept of information asymmetry play a crucial role in this scenario? Discuss how the parties involved could use Bayesian strategies to better navigate their decisions.

Section 10: Evolutionary Game Theory

Let's dive now into the world of Evolutionary Game Theory, a branch of mathematics that blends traditional game theory with evolutionary biology. This journey from conventional game theory concepts to an evolutionary perspective offers profound insights into how behaviors and strategies evolve over time.

Introduction to Evolutionary Game Theory

From Traditional to Evolutionary: The Evolution of Game Theory

Game theory originally focused on strategic interactions where the outcomes depend on the choices of all participants, often referred to as players. Traditional game theory analyses situations of conflict and cooperation through mathematical models. Think of it as a way to predict outcomes in scenarios where decision-makers interact, like businesses competing in a market or players in a poker game.

However, as insightful as traditional game theory is, it initially overlooked one crucial aspect—natural evolution and adaptation. This is where evolutionary game theory comes in, transforming the static approach of traditional game theory into a dynamic exploration of strategies in living populations.

This discipline has its roots in biology, sparked by researchers' desires to decode behaviors and evolutionary stabilities in animals. Pioneers like John Maynard Smith and George R. Price introduced the concept of an evolutionarily stable strategy (ESS) in the 1970s. An ESS is a strategy that, once prevalent in a population, cannot be supplanted by any rare alternative strategy. This pivotal concept will be your guide as you navigate through the intricacies of this field.

Evolutionary game theory examines how strategic interactions evolve over time, initially in biological contexts. Instead of focusing solely on human decision-makers, this theory considers how biological traits, viewed as strategies, pass from one generation to the next. These traits are not just physical but can also be behavioral patterns shaped by and shaping the evolutionary fitness of organisms.

The central premise of evolutionary game theory is not just to study strategies in a fixed setting but to see how they fare over repeated encounters and generations. It looks at which strategies are successful enough to be passed on to successive generations and how this affects the population as a whole.

Evolutionary Game Theory helps explain a variety of phenomena, such as the development of social norms, the evolution of cooperative behaviors, and the nature of conflicts and resolutions in different species, including humans. It provides a framework to analyze how certain behaviors or strategies become dominant, how they contribute to the stability of societies or groups, and how competing strategies interact with each other in the evolutionary game.

Evolutionary game theory enriches our understanding of strategic interactions from a long-term, dynamic perspective. It bridges the gap between static analysis and the fluid, ever-changing realities of natural and social behaviors. As we delve deeper into its core concepts in the following sections, we'll explore how strategies are not just chosen but evolve, and how this evolution shapes and is shaped by the organisms' interactions within their environments.

Core Concepts in Evolutionary Game Theory

Strategies as Behaviors That Pass Through Generations

In evolutionary game theory, strategies are more than just choices made in a game; they represent inherited behavioral traits that can be passed down through generations. These traits are subject to the forces of natural selection, much like physical characteristics in organisms. The survival and proliferation of these strategies depend not only on the outcomes they produce but also on how well they perform relative to other competing strategies in the environment.

Genetic Transmission and Cultural Transmission: Strategies can be transmitted genetically, where behaviors are encoded in the genes, or culturally, through learning and imitation. This dual mode of transmission allows for a rich tapestry of behaviors that can evolve over time. For instance, certain aggressive or cooperative behaviors in animals might be genetically determined, while in humans, much behavior is learned and thus transmitted culturally.

Phenotypic Plasticity: This concept refers to the ability of an organism to change its behavior in response to environmental conditions, demonstrating that strategies can be both inherited and modified. Phenotypic plasticity allows for adaptive behaviors that can respond to immediate environmental challenges, enhancing survival and reproductive success.

Evolutionary Stable Strategies (ESS)

An evolutionary stable strategy (ESS) is a strategy that, if adopted by a population in a certain environment, cannot be invaded by any alternative strategy that is initially rare. An ESS is conceptually similar to a Nash Equilibrium in traditional game theory, but it focuses on stability over time in the face of evolutionary pressures.

Definition and Implications: A strategy is evolutionary stable if, when most members of the population adopt it, there is no mutant strategy that could invade and outcompete it under the dynamics of natural selection. This concept helps explain why certain behaviors, such as specific mating rituals or aggression levels, persist within populations.

Examples of ESS: In the animal kingdom, the territorial behaviors of many species can be seen as ESSs. For example, when a particular type of aggressive behavior towards intruders secures a territory and access to resources, it likely becomes an ESS because any mutant strategy that is less aggressive may not be able to compete effectively for those resources.

The concept of "Fitness"

In the context of evolutionary biology and game theory, "fitness" refers to an organism's ability to survive and reproduce in its environment. It encapsulates how well an organism is adapted to the prevailing environmental conditions and how effective it is at passing its genes to the next generation. Essentially, fitness is a measure of reproductive success. Fitness can be understood through several key components:

- **Survival**: Before an organism can reproduce, it must survive to reproductive age. Therefore, traits that enhance an organism's ability to avoid predators, find food, and withstand environmental stresses contribute to its survival fitness.
- **Fecundity**: This refers to the reproductive rate of an organism—how many offspring it can produce during its lifetime. Traits that increase the number or viability of offspring directly contribute to an organism's fecundity.
- **Mating Success**: In many species, individuals must find and secure mates. Traits that enhance an organism's attractiveness to potential mates or its ability to compete with others for mates affect its mating success.
- **Offspring Viability**: Beyond just producing offspring, the fitness of an organism also depends on the viability of its offspring—their ability to survive and eventually reproduce themselves. This includes genetic health, the ability to acquire resources, and resistance to diseases.

Fitness is often measured relatively, comparing the reproductive success of different genotypes or phenotypes within a population:

- **Absolute Fitness** measures the ratio of the number of offspring (or genes) contributed to the next generation by a genotype relative to the number from the previous generation.

- **Relative Fitness** compares this ratio among different genotypes in the environment, often normalizing the most successful genotype as having a fitness of 1, with others expressed as fractions of this.

In evolutionary game theory, fitness is not just about physical survival and reproduction but can also be considered in terms of payoffs from different strategies in a game-theoretical context. Here, fitness could be represented by any measure of success that affects an individual's ability to pass on its genes or strategy, such as:

- **Gaining Resources**: In many games, such as the Hawk-Dove Game, payoffs can represent access to resources that enhance the individual's survival and reproductive success.
- **Social Standing or Territory**: In some models, the success in maintaining social standing or territory directly impacts reproductive opportunities and thus fitness.

Fitness in Business: Competitive Advantage

In business, fitness can be equated to a company's competitive advantage—its ability to outperform competitors due to superior capabilities, resources, or strategies. Here's how fitness manifests in a business context:

- **Market Share**: Like survival in nature, survival in business often hinges on capturing and maintaining a sufficient market share. A business that is 'fit' in its market environment effectively attracts more customers and defends against competitors, ensuring its continued existence and profitability.
- **Innovation**: Just as genetic mutations allow organisms to adapt to their environment, innovation allows companies to adapt to changing market conditions. A business that consistently innovates—whether through new products, improved processes, or advanced technologies—enhances its fitness by being better equipped to meet consumer needs and stay ahead of competitors.
- **Operational Efficiency**: Fitness can also be seen in how efficiently a company uses its resources. Higher operational efficiency, like higher energy efficiency in an organism, leads to better outcomes (profits) using fewer resources, enhancing the company's sustainability and ability to invest in growth.
- **Brand Strength and Reputation**: In the business ecosystem, a strong brand can be likened to advantageous genetic traits. It increases a company's fitness by fostering customer loyalty, allowing premium pricing, and providing a competitive edge that can be critical during economic downturns.

Fitness in Economics: Optimal Strategies

In economics, fitness can be associated with the effectiveness of strategies or policies in achieving economic goals such as growth, stability, and equitable distribution of resources.

- **Monetary Policy**: The fitness of different monetary policies can be assessed based on their success in controlling inflation and fostering conditions conducive to economic growth. For example, a policy that adeptly manages economic cycles demonstrates high fitness by maintaining stability and encouraging investment.
- **Fiscal Fitness**: Governments exert fiscal fitness through effective budget management. Policies that maximize fiscal fitness ensure optimal allocation of resources, sustainable debt levels, and funding for essential services, which in turn supports economic stability and growth.
- **Adaptation to Globalization**: Economic strategies that enable a country to adapt to and benefit from globalization reflect high fitness. This includes policies that foster competitiveness in global markets, such as investing in education and technology, improving infrastructure, and maintaining favorable trade relations.

Real-World Example: Fitness in Retail Business

Let's consider a real-world example of a retail company:

- **Company A** excels in supply chain management, allowing it to offer products at competitive prices quickly. It invests heavily in data analytics to understand consumer trends and adapts its inventory accordingly.
- **Company B**, on the other hand, struggles with inventory management and fails to leverage consumer data, resulting in frequent stockouts or overstock situations.

In this scenario, **Company A** demonstrates higher fitness by effectively aligning its strategies (supply chain management and data analytics) with the market environment. This alignment results in better customer satisfaction, increased sales, and higher market share, enhancing its survival and growth prospects compared to **Company B**.

Fitness in Social Contexts

Social fitness refers to the effectiveness of social behaviors or norms in enhancing the cohesion, well-being, and functioning of a community or society. This concept can be applied to various social phenomena, from the spread of cultural practices to the adoption of public policies. The components of Social Fitness are:

- **Cohesion and Stability**: Social behaviors or norms that promote group cohesion and social stability tend to have higher social fitness. For example, norms that encourage cooperative behavior, such as reciprocity or altruism, can increase the overall well-being of a community, making these norms more likely to be sustained and passed down through generations.
- **Adaptation to Social Changes**: Social practices that can adapt to changes—such as demographic shifts, technological advancements, or economic conditions—demonstrate high social fitness. For instance, educational systems that evolve to incorporate new technologies and teaching methods are more effective and remain relevant over time.
- **Enhancement of Social Capital**: Behaviors and norms that enhance social capital—meaning the networks of relationships among people who live and work in a particular society, enabling that society to function effectively—are also seen as having high social fitness. Trust, shared values, and participation in community activities are examples of social capital components.

Examples of Social Fitness are:

Effective Communication Norms: In any culture, norms that facilitate clear and effective communication tend to enhance social interactions and relationships. These might include norms around politeness, listening skills, or conflict resolution, which help maintain social harmony and are therefore socially fit.

Environmental Practices: Social norms that promote sustainability and environmental stewardship can gain fitness as the global emphasis on environmental health grows. Practices such as recycling, conservation, and the communal management of natural resources can help societies adapt to environmental challenges, thus enhancing their social fitness.

Health-Related Behaviors: Public health measures that become widely adopted, such as hygiene practices, are examples of behaviors with high social fitness. These behaviors directly contribute to the health and longevity of the population, increasing their propagation across the community.

The fitness of social strategies or norms can be measured by their prevalence, longevity, and the benefits they provide to the society. Metrics might include:

- Survey Data: Researchers can measure the acceptance and prevalence of certain norms through surveys that assess attitudes, behaviors, and values within a population.
- Societal Outcomes: The impact of social norms on societal outcomes like health statistics, crime rates, or educational achievements can indicate their fitness.
- Cultural Transmission: The extent to which cultural practices or norms are taught and maintained within families and educational systems also reflects their social fitness.

the evolutionary dynamics that also shape cultural and societal evolution, mirroring the biological processes that drive the survival and success of species.

Understanding fitness is crucial because natural selection acts on variations in fitness. Genetic variations that confer higher fitness are more likely to be passed on to subsequent generations, leading to the adaptation of populations over time. This process, driven by the differential reproductive success of individuals, shapes the evolution of species and the development of complex behaviors and traits.

Fitness Landscapes and Adaptation

The concept of fitness landscapes offers a vivid metaphorical depiction of how populations evolve over time in response to changes in strategy. A fitness landscape is a graphical representation that shows how different genotypes or strategies correlate with reproductive success, or fitness.

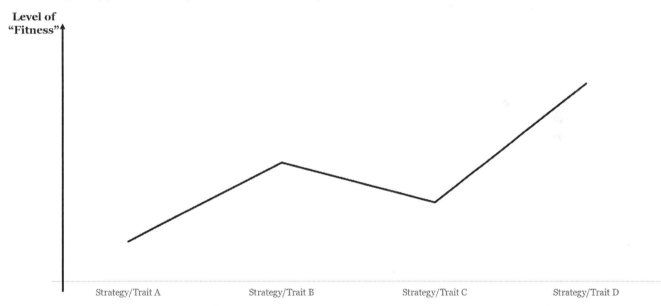

Navigating the Landscape: The landscape consists of peaks and valleys, where peaks represent high fitness and valleys represent low fitness. Organisms, through their strategies, 'move' over this landscape, generally evolving towards peaks of higher fitness. This movement is driven by mutation, selection, and genetic drift.

Adaptive Walks: As strategies evolve, populations perform what is known as an adaptive walk through the fitness landscape. This walk is the path a population takes as it adapts over time, generally moving towards local peaks in fitness. However, the landscape can change due to environmental shifts, leading populations to shift strategies and adapt to new peaks.

Real-World Applications: Understanding fitness landscapes is crucial in areas like antibiotic resistance for example, where bacteria evolve strategies to survive despite the presence of drugs designed to kill them. By mapping the fitness landscape of bacteria, researchers can predict potential paths of resistance evolution and design better treatment strategies.

The core concepts of evolutionary game theory—strategies as behaviors, evolutionary stable strategies, and fitness landscapes—provide profound insights into the dynamics of adaptation and survival in both biological and social systems. These ideas not only deepen our understanding of the natural world but also enhance our ability to predict and influence the evolution of behaviors in complex, interconnected environments. As we continue to explore these concepts, they will undoubtedly shed light on more facets of both human and non-human life, guiding future research and applications in diverse fields.

The Hawk-Dove Game

The Hawk-Dove Game is a conceptual framework used to analyze how aggression and cooperation manifest and evolve in different scenarios. It's based on the simple premise of two strategies—Hawk and Dove—competing for a shared resource. The names "Hawk" and "Dove" symbolize aggressive and peaceful strategies, respectively.

Strategies: Aggression vs. Cooperation

Hawk Strategy: Represents aggression. Hawks always fight to win the available resources. The outcome of this strategy depends on who the opponent is:

a) Against another Hawk: If a Hawk encounters another Hawk, they fight. This encounter results in a costly battle where both participants incur damage, though the winner gains the resource.

b) Against a Dove: If a Hawk encounters a Dove, the Hawk wins without a fight because the Dove retreats, avoiding conflict and thus incurring no injury.

Dove Strategy: Represents cooperation or conflict avoidance. Doves do not initiate fights and prefer to share or forfeit resources to avoid physical confrontations.

a) Against another Dove: If a Dove meets another Dove, they share the resource or incur minor costs due to a display of threat without actual fighting.

b) Against a Hawk: The Dove yields and walks away, leaving the resource to the Hawk. The Dove avoids the cost of fighting but gains nothing.

Outcomes and Why They Matter

The Hawk-Dove Game offers crucial insights into the balance between aggression and cooperation in a population:

- **Mixed Strategy Equilibrium**: Often, neither pure aggression nor pure cooperation is the most successful strategy by itself. The game typically reaches a mixed equilibrium where both Hawks and Doves exist in a population at specific proportions. This balance depends heavily on the cost of fighting relative to the value of the resource.
- **Evolutionary Stability**: The equilibrium is evolutionarily stable when no other strategy can enter the population and perform better than the existing mix of strategies. This stability helps explain why certain levels of aggression and cooperation are maintained in animal societies and human communities.

Practical Applications

Understanding the dynamics of the Hawk-Dove Game can also inform practical strategies in conservation efforts, business tactics, and policy-making:

- **Wildlife Management**: In managing wildlife and conservation areas, understanding the balance between aggressive and non-aggressive species can help in formulating strategies that protect endangered species while managing overpopulations.
- **Business Strategy**: Companies can analyze their competitive strategies in terms of Hawk and Dove, deciding when to aggressively pursue market share versus when to avoid costly competition.
- **Conflict Resolution**: In peacekeeping and conflict resolution, strategies from the Hawk-Dove Game can help mediators design interventions that promote cooperation and reduce hostility.

The Hawk-Dove Game not only serves as a fascinating model for studying conflict and cooperation but also provides a robust tool for predicting and managing the behaviors of individuals and groups in complex social and ecological systems. By exploring these dynamics, we can better understand the nature of interactions in both the animal kingdom and human society, leading to more effective strategies for managing conflict and fostering cooperation.

Replicator Dynamics

Replicator Dynamics is a fascinating concept from evolutionary game theory that helps explain how certain behaviors or strategies spread within a population over time. This framework provides a way to visualize and understand the evolution of these strategies based on their success relative to others within the same environment.

The Essence of Replicator Dynamics

At its core, Replicator Dynamics is about replication—how strategies that are more successful tend to be replicated more often. This replication can be biological, through reproduction, or cultural, through learning and imitation. The basic idea is that if a particular strategy (or behavior) offers an advantage—whether it's finding food, avoiding predators, or achieving social success—then individuals using this strategy are likely to have better outcomes. These individuals then pass on this successful strategy to the next generation either genetically or through social learning.

How Strategies Spread

Performance Relative to Others: In Replicator Dynamics, the key to a strategy's success is how it performs relative to other strategies within the population. If a strategy performs well, it becomes more common; if it performs poorly, it becomes less common. This dynamic is similar to natural selection in evolution, where advantageous traits become more prevalent over time.

Adaptation to the Environment: Strategies are not static. They evolve as the environmental conditions change. For example, a strategy that works well in a resource-rich environment might not perform as well in a resource-poor one. As environments shift, strategies that are better adapted to the new conditions will start to spread more widely.

Feedback Loops: The spread of strategies can create feedback loops. For instance, if a cooperative strategy begins to prevail in a population, it might change the social environment in ways that make cooperation even more beneficial, thereby accelerating the spread of cooperation.

Impact of Strategy Success on Population Dynamics

The success of a strategy affects not just the individuals who adopt it but the entire population. Here's how:

- **Population Structure**: As successful strategies spread, they can change the structure of the population. For instance, if aggressive strategies are successful, the population might become more aggressive overall.
- **Strategy Interaction**: Strategies interact in complex ways. The success of one strategy can depend heavily on the presence of other strategies within the population. For example, a strategy of deception might only be successful if most others are trustworthy.

Practical Implications

Understanding Replicator Dynamics has practical implications in various fields:

- **Conservation Biology**: By understanding which strategies are likely to succeed in a given environment, conservationists can better predict changes in species behaviors and populations under different ecological pressures.
- **Economics and Business**: In economics, companies might use principles similar to Replicator Dynamics to understand which business strategies spread under various market conditions and how they can adapt to remain competitive.
- **Social Policy**: Policymakers can use insights from Replicator Dynamics to craft interventions that promote beneficial social behaviors, such as cooperation and altruism, particularly in contexts like public health or community development.

Replicator Dynamics provides a powerful lens through which to view the evolution of behaviors and strategies in both biological and social systems. By focusing on how successful strategies replicate and spread, this approach helps explain the dynamic nature of adaptation and change in populations, offering valuable insights across a spectrum of disciplines.

Expanding the Horizons - Deeper Insights into Evolutionary Game Theory Applications

Economics: Market Evolution and Consumer Behavior

In economics, evolutionary game theory isn't just a theoretical construct; it directly impacts how markets evolve and how businesses adapt to changing consumer behaviors. Consider the organic food market, which has grown significantly as more consumers show a preference for healthier, environmentally friendly products. Businesses that adapt their strategies to include organic products can tap into this growing market segment. Over time, strategies promoting organic and sustainable practices might dominate as they align with increasing consumer awareness and regulatory pressures favoring sustainability.

Politics: Adapting Strategies in Political Campaigns

Political landscapes are particularly fertile grounds for the application of evolutionary game theory, where strategies must continuously evolve in response to shifting voter sentiments and political dynamics. For example, the use of social media in political campaigns has evolved dramatically. Initially, few politicians engaged with these platforms, but as their effectiveness became apparent, more politicians adopted these tools, and strategies for their use have continually adapted to include data analytics and targeted advertising to influence voter behavior more effectively.

Personal Life: Decision Making in Social Groups

In personal and everyday life, evolutionary game theory explains how individuals adapt their behaviors based on the actions and reactions of those around them. For instance, consider the evolution of social norms around recycling. As more people begin to recycle and share its importance, new community members or younger generations adopt these behaviors, making recycling a more dominant strategy within the community. This change happens as individuals see the benefits of recycling, supported by social approval and, increasingly, by municipal regulations.

Career: Evolution in Professional Environments

In career development, evolutionary game theory helps explain how certain professional strategies become predominant. In competitive job markets, for instance, the strategy of continuous skill enhancement through online courses and certifications has become increasingly prevalent. As more individuals adopt this strategy and succeed in securing advanced roles or promotions, others are likely to follow, reinforcing this approach as a dominant strategy in career development.

Business: Strategic Adaptations in Competitive Markets

Businesses often find themselves in a constant battle to outmaneuver competition and innovate to meet consumer demands. Consider the tech industry, where companies must rapidly evolve their product offerings and business strategies to stay relevant. Evolutionary game theory helps these companies understand which strategies (like investment in R&D or aggressive marketing) might lead to long-term dominance in the market.

Sports: Team Strategies and Performance Optimization

In sports, evolutionary game theory can be seen in how teams adapt their strategies based on past performances and opponents' tactics. Coaches might change their strategies for player rotations and game plays based on what has been successful in the past, which in turn influences how other teams prepare and counteract these strategies. This ongoing adaptation can lead to the evolution of playing styles within a league.

Technology: Security Strategies and Innovation Cycles

In technology, companies face the dual challenge of innovating rapidly while protecting their assets from cyber threats. Evolutionary game theory models how companies evolve their cybersecurity strategies in response to the changing tactics of cyber attackers. As new types of cyber threats emerge, companies adapt by developing and implementing more robust security measures, which then influences how hackers evolve their strategies, creating a continuous cycle of adaptation.

Social: Cultural Shifts and Behavior Evolution

Cultural norms and social behaviors also evolve under the influence of evolutionary game theory. As societal values shift, behaviors that were once rare—such as using renewable energy sources—can become common as individuals see the social and economic benefits of these strategies. Over time, these behaviors can become the norm, significantly influencing community practices and policies.

Environmental: Conservation Tactics and Sustainable Practices

Environmental strategies, particularly in conservation and resource management, often reflect evolutionary game theory. For example, strategies for water conservation evolve as conditions such as climate change and population growth exert pressure on water resources. Over time, water-saving practices that are initially adopted by a few can become widespread as their effectiveness becomes apparent and regulatory bodies begin to support these practices through incentives and legislation.

Legal: Negotiation Tactics and Evolutionary Legal Strategies

Finally, in the legal field, strategies for negotiations and litigation evolve as new laws and legal precedents are established. Lawyers adapt their approaches based on what strategies have been successful in the past and how legal standards evolve, which in turn influences how negotiations and litigation are approached in the future.

Reflection Questions

1) Exploring Evolutionary Strategies: Reflect on an industry or community that you are familiar with. How might evolutionary game theory explain the changes in strategies or behaviors observed over time within this group? Consider how adaptation to changing environments has influenced these strategies.

2) Applying Key Concepts: Think about a time when you had to adapt your strategy in response to changes in your personal or professional life. How does the concept of evolutionary stability, as discussed in evolutionary game theory, apply to your situation? What strategy proved to be stable, and why?

3) Real-World Insights: From the examples of evolutionary game theory applications provided in this section, choose one that particularly interests you. Analyze how the key concepts of evolutionary game theory are employed and discuss the implications of these strategies on the larger scale of the system or environment. How does this example broaden your understanding of how strategic interactions evolve over time?

Section 11: Mechanism Design - Shaping Decisions and Systems

Introduction to Mechanism Design

Mechanism design is a branch of economic theory that ingeniously works backward from desired outcomes to create systems that lead people to achieve specific goals. Unlike traditional game theory, which analyzes how players might act given certain rules, mechanism design involves setting up the rules themselves. It's about architecting the game so that even when players follow their own interests, the outcomes are beneficial for everyone involved. This section will ease you into the fundamental concepts of mechanism design, its significance, and the core principles that drive its application.

What is Mechanism Design?

Mechanism design is often described as reverse game theory. Imagine you're tasked not just with playing a game, but with designing it to ensure certain outcomes, regardless of how self-interested the players are. It's like being a social engineer, where your challenge is to align individual incentives with overall social good. The beauty of mechanism design lies in its ability to tackle complex problems across various fields—from auctioning off spectrum rights to designing climate policies or even structuring a company's internal policies to encourage productive behavior among employees.

The Core of Mechanism Design: Incentive Compatibility

A cornerstone concept in mechanism design is incentive compatibility. This principle ensures that every participant's best interest is to act in a way that truthfully reveals their private information or preferences. When you, as a designer, set up a mechanism, you're aiming to make sure that everyone involved is motivated to act honestly because doing so will lead them to the best possible outcome under the given rules.

Imagine you're organizing a charity auction. The goal is to raise as much money as possible. You design the auction rules to encourage bidders to bid their true value for the items on offer. If the auction is designed correctly, bidders reveal their true valuation through their bids, maximizing both their satisfaction and the charity's profits.

The Role of the Designer

As a mechanism designer, your role is complex and requires a deep understanding of both the goals to be achieved and the behaviors of those involved in the system. You need to anticipate the strategies that rational individuals might employ and craft rules that lead these strategies to produce desirable outcomes. This task often involves creative problem-solving and innovative thinking, as no one-size-fits-all solution exists for the diverse range of problems mechanism design can address.

Real-Life Impact of Mechanism Design

The applications of mechanism design are vast and varied. In the public sector, governments use mechanism design to structure taxes or allocate public goods efficiently and fairly. In the private sector, businesses use it to design contracts that align the interests of employers and employees or between firms and their suppliers.

For example, consider the problem of internet congestion. Internet service providers might use mechanism design to create pricing models that encourage users to moderate their usage during peak times, thus reducing overall congestion and improving service quality for everyone.

Fundamental Principles of Mechanism Design

The Essence of Mechanism Design

Mechanism design revolves around constructing frameworks within which individuals or organizations interact. Your goal as a designer is to create a system where, despite varying personal incentives, the outcome benefits all participants. This is achieved through carefully structured rules that ensure each player's choices lead to a desired collective result.

Incentive Compatibility: The Heartbeat of Mechanism Design

At the heart of effective mechanism design lies the principle of incentive compatibility. This crucial concept ensures that honesty pays off for all participants. When a mechanism is incentive compatible, every participant reveals their true preferences or information, as doing so maximizes their own benefit according to the mechanism's rules.

Real-Life Application: Imagine a local government plans to build a new community park. To fund this project, the government needs to determine how much each resident values the addition of the park, as this will influence both the design and scale of the project. The challenge lies in figuring out how much each citizen is willing to contribute financially.

In an ideal world, every citizen would honestly report how much the park is worth to them, ensuring that the funding accurately reflects the value it brings to the community. However, without a mechanism enforcing incentive compatibility, individuals might be tempted to underreport their true valuation to avoid a higher personal cost. This behavior leads to a shortfall in funding, which could either stall the project or result in a park that does not meet the community's needs or expectations.

To address this, the mechanism design must encourage truthful reporting. Here's where incentive compatibility comes into play. The government could design a funding mechanism that asks each citizen to report their valuation of the park and then contributes accordingly. However, to ensure that people report honestly, the mechanism includes a form of refund to align personal incentives with truthful reporting.

Contribution Based on Reported Valuations: Each citizen reports how much the park is worth to them, and initially, pays a share of the cost based on this amount.

Excess Funding and Refunds: Suppose the total collected exceeds what is needed (as the project cost is fixed). The mechanism then refunds the excess amount to each contributor in a way that the final cost paid by each aligns more closely with their true valuation.

For instance, if a resident reports the park is worth $100 to them but the excess funds allow for a $20 refund per resident, their net contribution aligns closer to their actual valuation, assuming others also report truthfully. This refund acts as an incentive for honest reporting from the outset because each citizen knows that over-reporting will lead to unnecessarily high initial payments, while under-reporting could result in receiving a smaller refund or even additional charges to meet the project costs.

This approach ensures that the project is adequately funded without overcharging anyone, aligning individual incentives with collective benefits. It leverages the principle of incentive compatibility effectively by making honesty the best policy for all involved. Residents feel more engaged and fairly treated, knowing that their contributions are directly tied to their personal valuations, and the project funding reflects the genuine value it offers to the community.

Individual Rationality: Ensuring Participation

For any mechanism to be effective, it must satisfy the condition of individual rationality. This means that participating in the mechanism must be at least as good for each player as not participating at all. In other words, the mechanism should offer each participant a payoff that meets or exceeds their utility from opting out, ensuring voluntary participation.

Example in Action: Imagine a company designing a bonus scheme to motivate employees. If the scheme provides bonuses based on achievable targets and the extra effort leads to higher personal income than not participating, employees are likely to engage enthusiastically, driving productivity up.

Social Choice and Efficiency: Balancing Acts

Mechanism design also deeply involves social choice theory, which deals with aggregating individual preferences or interests into a collective decision. The efficiency of a mechanism often relates to how well it can aggregate these preferences to achieve a socially desirable outcome.

Efficiency in Mechanisms: If we consider the example of auction design, for instance, the goal is often to maximize total welfare, meaning the items for sale should go to those who value them most highly. An efficiently designed auction ensures that each item is allocated in a way that maximizes total value derived from the item, which is a direct application of social choice theory ensuring efficient outcomes.

The Revelation Principle: Simplifying Complexity

The revelation principle is a powerful tool in mechanism design. It states that for any mechanism achieving a certain outcome, there is a truthful mechanism — one requiring participants to simply reveal their private information honestly — that can achieve the same outcome. This principle simplifies the design of mechanisms by focusing on those where players truthfully report their types.

Utilizing the Revelation Principle: In the example where a local government wants to decide whether to build a public park. By designing a survey mechanism where residents report how much they value the park, assuming they report truthfully, the government can easily decide whether the total benefit justifies the cost, simplifying decision-making.

Dynamic Mechanism Design: Adapting Over Time

As environments and individual preferences change, static mechanisms may fail to achieve the desired outcomes consistently. Dynamic mechanism design addresses this by allowing the rules to adapt based on observed behaviors and outcomes over time.

Dynamic Adaptations: In dynamic pricing models, for example, prices are adjusted based on demand fluctuations. Ride-sharing platforms like Uber use dynamic pricing mechanisms that increase prices during high demand periods, balancing supply and demand efficiently and ensuring service availability.

Exploring Types of Mechanisms in Mechanism Design

In mechanism design, understanding the various types of mechanisms is crucial for effectively addressing diverse strategic goals. Each type of mechanism, from auctions and voting systems to market and contract mechanisms, is tailored to solve specific problems by aligning individual incentives with desired outcomes. This exploration will guide you through these different mechanisms, illustrating how each functions and is applied in real-world scenarios, helping you grasp their utility and strategic value.

Auction Mechanisms

Auctions are one of the most prevalent and studied mechanisms in economic theory and practice. They are designed to allocate resources or goods based on competitive bidding, which ideally reveals the true value participants place on the item being auctioned.

How Auctions Work: In a typical auction, you, as a bidder, will compete by placing bids on an item you desire. The auction rules—whether it's an open ascending price auction (English auction), a descending price auction (Dutch auction), or a sealed-bid auction—define how bids are made and how the winner is determined. The goal is to design the auction to maximize efficiency, where the item goes to the highest valuer, and to optimize revenue for the seller.

Real-World Example: Consider the spectrum auctions conducted by governments to allocate radio frequencies to telecommunication companies. These auctions are critical because they ensure that a valuable public resource is allocated efficiently, preventing any single entity from monopolizing the spectrum without fair competition.

Voting Mechanisms

Voting systems are mechanisms designed to aggregate individual preferences into a collective decision. These mechanisms are fundamental in democratic societies and various organizations where group decision-making is required.

Understanding Voting Systems: In a voting mechanism, your vote contributes to the collective decision-making process, which could be about electing a representative, passing a law, or making a policy decision. The design of the voting system, whether it's a simple majority rule, a proportional system, or a ranked-choice voting system, impacts how effectively it translates individual preferences into a group decision.

Application in Everyday Governance: In political elections, mechanism design is used to create voting systems that reflect the will of the populace accurately and fairly, minimizing issues like vote splitting or the marginalization of minority opinions.

Market Mechanisms

Market mechanisms are designed to facilitate voluntary exchanges between buyers and sellers, with prices adjusting based on supply and demand. These mechanisms are integral to the functioning of economies, allowing resources to be allocated efficiently without central planning.

Market Mechanism Dynamics: In a market mechanism, like a stock market or a commodities exchange, you participate either as a buyer or a seller. The mechanism provides a platform where prices adjust dynamically based on the collective actions of all market participants, aligning individual buying and selling decisions with market equilibrium.

Illustration in Energy Markets: Consider electricity markets where utility companies need to ensure that electricity supply matches demand throughout the day. Market mechanisms allow for real-time pricing, which encourages consumers to use electricity during off-peak hours, aiding in demand management and grid stability.

Contract Mechanisms

Contract mechanisms are designed to ensure that agreements between parties are mutually beneficial and that each party has an incentive to fulfill their part of the agreement. These are crucial in scenarios where activities are interdependent, and cooperative behavior needs encouragement.

Mechanics of Contract Design: In a contract mechanism, terms are set so that the outcomes for all parties are optimized when everyone acts according to the agreement. This often involves stipulations that provide incentives for performance and penalties for non-compliance.

Example in Employment Contracts: In the corporate world, employment contracts often include performance bonuses linked to specific targets. This mechanism incentivizes employees to align their efforts with the company's goals, ensuring productivity and mutual profitability.

Designing Mechanisms for Specific Goals: Crafting Systems to Drive Desired Outcomes

In mechanism design, the fundamental challenge is to engineer environments that steer individual decisions towards achieving specific, often collective, objectives. Whether it's encouraging fair resource distribution, maximizing economic efficiency, or promoting public welfare, the mechanisms devised play a critical role in shaping the outcomes of various interactions. Here we will go into the nuances of designing mechanisms tailored to meet distinct goals, emphasizing practical applications across various sectors.

Maximizing Economic Efficiency through Auctions

Auctions are a cornerstone of economic mechanism design, used widely from treasury bills to online advertising spaces. Their primary goal is to allocate resources efficiently to those who value them most, thereby maximizing total economic welfare.

Spectrum Auctions for Telecommunications - Consider again the design of spectrum auctions where government bodies allocate frequency bands to telecom operators. The challenge is to design these auctions in a way that not only maximizes revenue but also ensures that the spectrum is used efficiently to benefit consumers. The sealed-bid second-price auction, or Vickrey auction, is often employed. In this model, each bidder submits a bid in a sealed envelope. The highest bidder wins but pays the price offered by the second-highest bid. This setup encourages bidders to reveal their true valuation of the spectrum, leading to an efficient allocation at a fair price, reflecting the true market value without inflating costs unnecessarily.

Enhancing Social Welfare with Sliding Scale Fees

When designing mechanisms for public services, such as healthcare or education, promoting accessibility and affordability is key. Sliding scale fee systems are an excellent example of how mechanism design can enhance social welfare by tailoring costs based on an individual's ability to pay.

Application in Community Health Clinics - In community health clinics, sliding scale fees adjust charges based on a patient's income and family size. This approach not only ensures that all individuals have access to necessary medical services regardless of their financial situation but also maximizes the utilization of healthcare resources. By allowing more people to afford preventive care, the system helps reduce the occurrence of severe health issues that would require more expensive treatments, thus saving costs for the entire healthcare system over time.

Promoting Fairness in Resource Allocation

Fairness is particularly critical in mechanisms that involve sharing or distributing scarce resources, whether these are educational opportunities, organ transplants, or financial aid.

College Admissions - When it comes to college admissions, the challenge lies in fairly distributing limited spots among a large pool of applicants. Each student has their own set of preferred colleges, and each college has its own criteria for selecting students. The goal is to match each student to a college in a way that is fair to all involved and stabilizes the admission process.

The stable *matching algorithm, developed by Gale and Shapley*, is a method used to ensure that the match between students and colleges is as fair and stable as possible. Here's how it simplifies and improves the college admissions process:

Mutual Preferences: Each student submits a list ranking their preferred colleges in order of desire. Similarly, each college ranks the students who have applied to them, based on criteria such as test scores, extracurricular activities, and interviews.

The Matching Process: The algorithm begins by having each student "apply" to their top-choice college according to their list. Each college then considers the students that have applied and tentatively accepts students up to the number of available spots, prioritizing them according to its own rankings. If more students apply than there are spots available, the least preferred students (according to the college's rankings) are rejected, but they are not out of the process yet.

Iteration and Adjustment: Rejected students proceed to apply to their next-choice college. This college then reviews its new applicants along with its tentatively accepted students. If it prefers some new applicants over its current ones, it may accept these new applicants and reject others. This process continues, with students who are rejected from one college applying to their next choice, until no student wants to move from their current placement.

Stability: The process results in a stable match, meaning there are no pairs of a college and a student who would prefer each other over the arrangements made by the algorithm. Every student is matched with a college such that there is no other college they would prefer more than the one they are matched with. This assumes that each student has ranked the colleges based on their preferences, from the most to the least desirable. Even if a student prefers another college to the one they are matched with, that alternative college must also prefer this particular student over at least one of the students it has already accepted. If the college would not prefer this student over its current students, then the match remains stable.

Imagine a scenario where we have two colleges, College A and College B, and two students, Student 1 and Student 2. Each student has a preference order for the colleges, and each college also ranks the students based on its admission criteria. Preferences:

- Student 1: College A > College B
- Student 2: College A > College B
- College A: Student 2 > Student 1
- College B: Student 1 > Student 2

If we match Student 1 with College B and Student 2 with College A:

- Student 1 is at College B but prefers College A. However, College A prefers Student 2 over Student 1.
- Student 2 is at College A, which is their top choice.

Even though Student 1 would be happier at College A, College A would not prefer to have Student 1 over Student 2, whom it currently has. Hence, no student is in a position to move to another college where they would be happier and that college would also prefer them over its current students.

This algorithm promotes fairness by ensuring that every student has a fair shot at being admitted to their preferred colleges based on their qualifications and preferences. It also enhances stability in the admissions process by ensuring that once matches are made, they are final—there are no incentives for students or colleges to break the agreement and seek a different arrangement. This stability is crucial because it prevents a chaotic situation where students and colleges continually attempt to rearrange placements.

Encouraging Environmental Sustainability

Environmental challenges require mechanisms that promote sustainability through conservation and efficient resource use. Cap-and-trade systems for carbon emissions exemplify how mechanism design can drive environmentally friendly behaviors.

Carbon Emissions Trading - In a cap-and-trade system, a government sets a cap on the total amount of greenhouse gases that can be emitted. Companies receive or buy emission permits, which they can trade with each other. As companies implement greener technologies, they reduce their emissions, allowing them to sell excess permits. This market-based approach incentivizes companies to innovate and reduce emissions by providing a financial reward for doing so, aligning individual profit motives with broader environmental goals.

Challenges and Limitations in Mechanism Design

Mechanism design is a powerful tool in economics and social sciences, providing structured solutions to complex problems by aligning individual incentives with desired outcomes. However, like any sophisticated tool, it faces its own set of challenges and limitations. Understanding these hurdles is crucial for you as it prepares to anticipate potential pitfalls and consider necessary adjustments In the approach.

Information Asymmetry

One of the fundamental challenges in mechanism design is dealing with information asymmetry, where different participants in a mechanism have access to different amounts of information. This discrepancy can lead to suboptimal outcomes if not properly managed.

Illustration: Imagine a scenario where a government is trying to allocate funds to different regions based on need. If the regions do not accurately report their needs or if the government cannot verify these reports, the allocation mechanism may fail to distribute the funds in the most effective manner. This could result in some regions receiving more than they need while others suffer due to a lack of resources.

Incentive Compatibility

Creating mechanisms where every participant's best strategy is to follow the rules as intended—known as incentive compatibility—is immensely challenging. Incentive compatibility is crucial for ensuring that the mechanism functions as designed, but aligning these incentives with the overall goals of the mechanism can be intricate and difficult.

Example: Consider a carbon trading system designed to reduce emissions. If the penalties for exceeding emission caps are too low, companies might find it cheaper to pay the penalties rather than invest in cleaner technology. This misalignment of incentives can lead to the mechanism failing to achieve its environmental goals.

Complexity and Cost of Implementation

The design and implementation of mechanisms often involve complex calculations, negotiations, and enforcement measures, all of which can be costly and time-consuming.

Real-World Challenge: Implementing a nationwide healthcare enrollment system that allows people to select from various providers and plans can be exceedingly complex. The mechanism must account for millions of individuals with varying health needs and financial capabilities, and creating a system that is both fair and efficient requires significant administrative oversight and technological infrastructure.

Limited Predictability and Unintended Consequences

Mechanisms are generally designed based on assumptions about human behavior, which may not always hold true in practice. Unanticipated behaviors can lead to unintended consequences, sometimes making a well-intentioned mechanism counterproductive.

Scenario: A city introduces a traffic congestion pricing mechanism, charging drivers a fee to enter the busiest parts of the city during peak hours. While this might reduce traffic as intended, it could also lead to increased congestion in neighboring areas not subject to the charge, or disproportionately affect low-income drivers who cannot afford to pay the fee but must travel to these areas for work.

Resistance to Change and Adoption

Introducing a new mechanism often faces resistance from those who benefit from the status quo or who fear losing out from the change. Overcoming this resistance and garnering enough support to implement a new mechanism can be a significant challenge.

Situation: Rolling out a new digital voting system to replace traditional paper ballots may meet resistance from older voters who are not comfortable with technology, as well as from political groups that benefit from lower turnout among demographics less likely to use digital tools.

Ethical Considerations and Fairness

Mechanism design must also navigate the delicate balance of ethical considerations and fairness, ensuring that no group is unduly disadvantaged by a mechanism.

Ethical Dilemma: A mechanism designed to allocate scarce medical resources, like organs for transplant, must consider not only the medical urgency and likelihood of success but also ethical dimensions such as patient age, quality of life post-transplant, and equity of access regardless of socio-economic status.

Advanced Topics in Mechanism Design

Dynamic Mechanisms: Adapting to Changing Environments

In dynamic mechanisms, the focus is on designing systems that are not only effective at a single point in time but also adapt to changing circumstances. Dynamic mechanisms are crucial in environments where information evolves or where participant behaviors change over time.

Understanding Dynamic Mechanisms: Dynamic mechanisms adjust their rules or outputs in response to new data or outcomes. For instance, a dynamic pricing mechanism in the electricity market may change prices minute by minute, based on fluctuating supply and demand to ensure efficiency and stability in the grid.

Application in Renewable Energy Markets: Consider the integration of renewable energy sources like solar and wind, which are intermittent and unpredictable. Dynamic mechanisms that adjust energy prices in real-time can incentivize consumers to use energy during off-peak times, helping to balance the grid and reduce reliance on non-renewable energy sources.

Interdependent Values: Complex Valuation in Auctions

Interdependent values occur when the value of an item to one participant depends on the private information or valuations of other participants. This complexity is particularly prevalent in auctions for artworks, antiques, or mineral rights, where the true value of the item is uncertain and may depend on external factors or assessments.

Exploring Auctions with Interdependent Values: In such auctions, traditional bidding strategies may lead to the winner's curse, where the winner ends up overpaying. Mechanism designers must create auction formats that mitigate these risks, encouraging information sharing or designing rules that adjust to revealed valuations dynamically.

Example with Oil Drilling Rights: Auctions for oil drilling rights are a classic case where interdependent values are significant. Bidders might rely on geological surveys and the bids of other participants to gauge the value of drilling rights. Mechanisms such as common-value auctions, which allow bidders to adjust their bids as new information is revealed, can help in such settings.

Leveraging Blockchain for Transparent Mechanisms

Blockchain technology offers unprecedented opportunities to design mechanisms that are transparent, tamper-proof, and decentralized. This is particularly beneficial in contexts requiring high levels of trust and verifiability.

Blockchain in Voting Systems: Implementing blockchain technology can revolutionize voting mechanisms, ensuring that votes are recorded accurately and remain unchangeable once cast. This system enhances transparency and trust, reducing the risk of fraud and manipulation in elections.

Smart Contracts in Mechanism Design: Smart contracts on blockchain platforms enable the automatic execution of agreements based on predefined rules. For instance, an insurance payout mechanism could automatically trigger payments to policyholders based on verifiable data inputs, such as weather conditions meeting the criteria for a flood insurance claim.

Artificial Intelligence in Mechanism Design

Artificial intelligence (AI) transforms mechanism design by enabling the analysis of large datasets and the modeling of complex decision environments. AI can predict outcomes, optimize mechanism parameters, and simulate the behavior of participants with unprecedented precision.

In financial markets, AI can help design mechanisms that predict market trends and react in real-time to changes. AI algorithms could manage buying and selling orders in stock exchanges to maximize efficiency and minimize market impact.

AI technologies allow also for the personalization of mechanisms based on individual behaviors and preferences. For example, personalized pricing mechanisms in e-commerce can offer dynamic pricing based on a customer's browsing and purchasing history, optimizing sales and customer satisfaction.

The Future of Mechanism Design

As we look toward the future, the fields of mechanism design are set to expand significantly. The integration of advanced technologies like AI and blockchain, along with deeper insights into dynamic and complex interactions, will open up new possibilities for designing mechanisms that are more efficient, fair, and responsive to human needs. These advancements promise not only to solve traditional problems more effectively but also to tackle new challenges that arise in an increasingly interconnected and technologically sophisticated world. As you engage with these advanced topics, consider how they might apply to your own areas of interest or professional endeavors, offering tools to craft innovative solutions to the complex strategic challenges of tomorrow.

Exploring Other Real-World Applications of Mechanism Design

Mechanism has practical implications across a diverse array of fields. Whether you're navigating the complexities of economic markets, making policy decisions, or simply engaging in everyday interactions, understanding how mechanism design is applied can provide you with valuable insights and tools. Here, we delve deeper into how mechanism design influences various aspects of life and society, offering real-world applications that demonstrate its versatility and power.

Politics: Public Decision Making

Mechanism design is crucial in decisions related to public goods allocation, like funding for public projects. Mechanisms can be designed to ensure that the distribution of funds for public goods like parks, roads, and schools reflects the needs and values of the community, often through participatory budgeting processes where community members have direct say in spending decisions.

Personal Life and Everyday Interactions

Even in personal life and daily interactions, mechanism design has its applications. Consider a family deciding on a vacation destination or a group of roommates determining how to divide household chores and expenses. Mechanism design can help create fair systems where each person's preferences and inputs are considered, leading to outcomes that everyone can agree on.

Negotiating Personal Relationships

In personal relationships, mechanisms like bargaining and fair division algorithms help resolve conflicts by ensuring that agreements are fair and satisfactory to all parties involved. For example, dividing possessions during a move or a breakup can be facilitated by algorithms that ensure each person feels they have received a fair share.

Career: Job Matching and Performance Incentives

In career development, mechanism design is used to create systems that match employees with jobs in a manner that maximizes satisfaction and productivity. Platforms like LinkedIn use sophisticated algorithms to match job seekers with positions that suit their skills and career goals, benefiting both employers and employees.

Performance Incentives

Companies often design performance incentives, such as bonuses and stock options, to align employees' goals with company objectives. Effective mechanism design ensures that these incentives motivate desired behaviors, like innovation and teamwork, leading to better outcomes for the company and satisfying careers for employees.

Business: Competitive Strategies and Contract Design

In the business world, companies use mechanism design to develop competitive strategies and construct contracts that mitigate risks and align incentives between parties. For instance, supply chain contracts are designed to ensure that suppliers deliver goods on time, in the right quantity, and at agreed-upon quality levels, often including penalties for non-compliance and rewards for exceeding targets.

Sport: Game Theory and Strategy Development

In sports, mechanism design is used to develop game strategies and rules that promote fair play and competition. The drafting rules in professional sports leagues are designed to maintain competitive balance by allowing weaker teams to pick first from available new talent.

Reward Systems in Sports

Sports leagues also use mechanisms to determine payouts for wins and other achievements, designed to reward performance and encourage competition. These systems need to balance rewarding success and maintaining financial sustainability.

Social: Community Engagement and Resource Sharing

Mechanism design fosters effective community engagement and resource-sharing initiatives, from carpooling platforms to community-supported agriculture (CSA) programs. These mechanisms need to balance individual preferences with community benefits, ensuring that resources like food from local farms or rides in a carpool are distributed in ways that maximize social welfare and environmental benefits.

Social Networking and Content Moderation

Social media platforms use mechanisms to moderate content and manage user interactions to foster positive community interactions and prevent harmful behaviors. The design of these mechanisms needs to encourage active, positive participation while minimizing the spread of misinformation and toxic behavior.

Legal: Dispute Resolution and Regulatory Compliance

In the legal field, mechanism design helps create systems for dispute resolution that are fair and efficient, such as arbitration and mediation mechanisms that provide alternatives to traditional courtroom battles. These systems are designed to resolve disputes by aligning the interests of all parties to achieve mutually acceptable outcomes without the need for lengthy and costly litigation.

Regulatory Compliance Mechanisms

Regulatory mechanisms are designed to ensure that companies comply with laws and regulations, from environmental standards to financial reporting requirements. These mechanisms often include audits, reporting requirements, and penalties for non-compliance, structured to incentivize companies to adhere to legal standards while promoting transparency and accountability.

Technology-Driven Future: AI and Advanced Analytics

Looking ahead, the integration of AI and advanced analytics promises to further refine mechanism design, enabling more precise modeling of complex systems and better prediction of outcomes. AI can help simulate and test different mechanism designs before they are implemented, reducing the risk of unintended consequences and allowing for fine-tuning based on predicted behaviors.

AI in Personalized Medicine

In the medical field, AI is revolutionizing mechanism design by personalizing treatment plans based on genetic information. This approach uses mechanisms that optimize treatment protocols based on individual patient data, improving outcomes and reducing side effects. AI models process vast amounts of data to design treatment mechanisms that are highly tailored to individual needs, pushing the boundaries of what's possible in medical treatment.

Reflection Questions

1) Understanding Mechanism Design: Reflect on a system or process in your workplace or a community organization where the outcomes depend significantly on the participants' decisions. How could the principles of mechanism design be applied to redesign this system for more effective outcomes? What specific mechanisms might you implement?

2) Evaluating Mechanism Types: Consider a personal or professional scenario where you could design a mechanism to achieve a desired outcome. What type of mechanism would be most appropriate (e.g., auction mechanisms, voting systems, market mechanisms) and why? How would you ensure that the mechanism aligns with the fundamental principles of mechanism design?

3) Real-World Application Challenges: Choose a real-world application of mechanism design from this section. Discuss the challenges and limitations that might arise when implementing such mechanisms. What strategies could be employed to overcome these challenges? How do these considerations impact the effectiveness of the designed mechanism?

Part 4 - Real Life and Concrete Applications of Game Theory

Section 12: Game theory in personal life and everyday decisions.

Family Decision-Making and Household Dynamics

In every family or household, decisions need to be made that affect all members, such as budgeting, holiday planning, or even deciding on meal plans. Game theory can help in these situations by offering strategies that ensure cooperation and fair division of resources.

Scenario: Allocating Chores Equitably - Imagine your household trying to divide up chores. Each person has chores they prefer and those they dislike. Using game theory, you can implement a system where each person ranks their preferences, and chores are allocated based on these rankings to maximize overall happiness. This can prevent disputes and ensure that chores are done efficiently and without resentment.

Financial Decisions: Saving, Investing, and Purchasing

Financial decisions are often complex and involve weighing various risks and rewards, a perfect situation for applying game theory.

Example: Planning a Family Budget - When planning a budget, each family member may have different priorities and spending habits. Game theory can help you negotiate these priorities and come up with a budget that maximizes overall utility or satisfaction for the family. This involves identifying common goals, potential trade-offs, and strategies that encourage saving and wise spending.

Social Interactions and Event Planning

Organizing events or even just deciding where to meet friends can involve strategic decision-making. Game theory helps in understanding and balancing different preferences to optimize collective enjoyment.

Planning a Gathering - When planning a social event, like a dinner out with friends, game theory can assist in choosing a restaurant. If everyone lists their top choices, you can use a simple voting mechanism to pick a location that most satisfies the group's preferences, avoiding conflicts and dissatisfaction.

Dating and Relationship Dynamics

Game theory isn't just about competitions or economic transactions—it also applies to dating and relationships, where understanding and managing expectations and behaviors are crucial.

Navigating First Dates On a first date, both parties might play a 'game' to try and figure out the level of interest of the other. Each person decides how much about themselves to reveal and how to interpret the signals from the other person. Game theory can offer insights into the best strategies for revealing your own interests and intentions while also correctly interpreting the other person's actions and words.

Negotiating Personal Conflicts

Conflicts are inevitable in any relationship, whether with family, friends, or partners. Game theory provides tools to approach conflict resolution strategically, aiming for outcomes that preserve relationships and lead to mutually agreeable solutions.

Resolving Disputes - In a dispute, such as deciding who gets to use a shared vehicle at a particular time, game theory can suggest compromise solutions or trade-offs. For example, if two parties value different aspects of usage (time vs. days), they can negotiate based on these preferences to reach a satisfactory arrangement without conflict.

Personal Growth and Learning

Game theory also applies to personal development, particularly in situations where you need to decide how much time and resources to invest in your own skills and knowledge.

Investing in Skills - Deciding whether to learn a new skill or language involves evaluating the potential benefits against the time and effort required. Game theory can help you analyze these decisions, especially when considering the competitive advantage these skills might provide in your career or personal life, allowing you to make strategic choices about your personal development.

Section 13: Economics and Market Analysis Through the Lens of Game Theory

In the world of economics, game theory provides powerful insights that help explain and predict the behavior of various market participants. Whether you're examining competition between corporations, pricing strategies, or market behavior under different economic policies, game theory is an indispensable tool in your analytical arsenal. Here, we delve into several aspects of economics where game theory not only clarifies complex dynamics but also guides strategic decision-making, offering a practical perspective on its applications.

Understanding Oligopolies and Competitive Dynamics

One of the classic applications of game theory is in analyzing oligopolies—market structures with a small number of firms. In such markets, the actions of one company can significantly affect the entire market, leading to a complex web of strategic decisions.

Case Study: Price Wars and Collusions - Imagine you are a key decision-maker in a company that operates within an oligopoly. Each firm in this market sets prices, not just based on costs and profit margins, but also on expected reactions from competitors. Game theory models, such as the Cournot model and the Bertrand model, help predict whether competitors will match price cuts or engage in non-price competition. For instance, if two companies continuously undercut each other's prices in a bid to gain market share, it could lead to a price war, reducing profits for all players in the market. Game theory provides frameworks like Nash Equilibrium, helping you understand under what conditions firms might instead tacitly agree to maintain higher prices, maximizing collective profits without explicit collusion, which is illegal in many regions.

Auctions and Bidding Strategies

Auctions are another area where game theory is extensively applied. From government bonds to spectrum sales, auctions are fundamental in allocating resources and contracts efficiently.

Exploring Auction Formats - Different auction formats can lead to dramatically different outcomes. For example, in a sealed-bid auction, you might bid aggressively to outperform your rivals, knowing that the highest bid wins, but without knowing the value others place on the same item. In contrast, an English auction, with its ascending price and public bids, allows bidders to react to each other's moves, potentially leading to higher bids as competitors adjust their valuations based on visible information. Game theory helps auction designers and participants predict and strategize under different auction rules, maximizing their outcomes based on expected behaviors of others.

Market Design and Matching Theory

Market design, a field closely related to game theory, involves creating mechanisms that allow for efficient matching of buyers and sellers, students to schools, or even organ donors to recipients. Matching theory considers preferences of all parties to ensure that the matches are stable—no two agents would prefer to be matched with each other over their current matches.

Application in School Admissions In many cities around the world, students and schools are matched using algorithms derived from game theory principles. These algorithms consider the preferences of both students for schools and schools for students, aiming to place as many students as possible into their top-choice schools while ensuring that schools also receive candidates they prefer. This not only improves satisfaction among students and parents but also enhances educational outcomes by aligning capabilities and educational goals.

Behavioral Economics and Game Theory

Behavioral economics, which examines the effects of psychological, cognitive, emotional, cultural, and social factors on decisions, often uses game theory to analyze decisions that deviate from traditional economic predictions. This includes understanding how people might irrationally weigh risks and rewards or how trust and fairness considerations might override straightforward profit maximization.

Exploring Public Goods and Free-Rider Problems Consider a situation involving public goods, like a community project to build a park. Game theory and behavioral economics together assess how individuals might contribute to a public good voluntarily. Despite the free-rider problem—where individuals might benefit from the good without contributing to its costs—experimental games show that many people are willing to contribute more than what purely self-interested motives would dictate. This can be crucial for public policy and fundraising strategies, indicating that under certain conditions, appealing to community spirit and fairness can be very effective.

Financial Markets and Strategic Trading

In financial markets, traders and firms often use game theoretic strategies to maximize their gains. This includes anticipating moves of other market participants, strategizing entry and exit times, or manipulating market perceptions through strategic trades.

High-Frequency Trading - In high-frequency trading, firms use algorithms that can execute trades in milliseconds to gain an advantage over competitors. Game theory models help in designing these algorithms so that they can predict and react to patterns in order placements and cancellations, essentially 'playing' a high-speed game against other algorithms to capitalize on small price changes before they disappear.

Corporate Strategy and Negotiation

Corporate strategy often involves game-theoretic thinking, especially in negotiation settings where companies must anticipate and counteract the moves of their competitors, suppliers, or even regulators.

Mergers and Acquisitions During mergers and acquisitions, for instance, both the buying and selling firms engage in a game of incomplete information, each trying to ascertain the true value of the deal and the intentions of the other side without revealing too much. Game theory provides tools to strategize these interactions, determining when to push for a better deal, when to walk away, or when to seek alternative partnerships.

Exploring the Power of Game Theory in Advertising and Marketing

Marketing strategies often resemble games, especially in competitive sectors where firms must not only attract consumers but also outmaneuver competitors. Game theory helps in designing promotional campaigns and pricing strategies that anticipate competitive responses, ensuring that a firm's marketing moves cannot be easily neutralized.

Dynamic Pricing Strategies Dynamic pricing is a strategy where prices are adjusted based on real-time demand and supply conditions, competitor prices, and other market dynamics. Game theory models can predict how competitors might respond to changes in your pricing, and what that means for consumer demand, helping you optimize your pricing strategy not just for immediate profit but for long-term competitive advantage.

Section 14: Political Science and Game Theory

Electoral Strategies and Voting Systems

In electoral politics, game theory plays a crucial role in shaping the strategies of candidates and political parties. Each player in this political game—the candidates—aims to maximize their chance of winning by choosing positions and strategies that appeal to the broadest possible electorate.

Campaigning Tactics - Imagine you are a political strategist. Using game theory, you can determine which issues to emphasize in a campaign to appeal to swing voters without alienating the base. This involves modeling voter preferences and predicting opponent moves, effectively playing a game of positioning and counter-positioning based on anticipated voter reactions.

Designing Voting Systems - Game theory also aids in designing voting systems that reflect the electorate's true preferences. For instance, systems like ranked-choice voting can reduce the occurrence of spoiler effects, where a third candidate might split the vote, often leading to the election of a less popular candidate. This method encourages voters to rank candidates by preference, ensuring that their votes contribute to electing a broadly acceptable candidate, even if their top choice cannot win.

Legislative Bargaining and Coalition Building

In legislative bodies, whether national parliaments or local councils, game theory can be used to analyze how coalitions are formed and how legislative bargains are struck.

Coalition Formation - Consider a situation where no single party has an outright majority. Game theory models can help in understanding how parties negotiate to form coalitions, aligning their policy goals sufficiently to govern. These models assess the stability of potential coalitions and the likely concessions each party must make, facilitating more strategic negotiation processes.

Amendment Games in Legislation - When a piece of legislation is on the floor, different members or parties can propose amendments. Game theory helps in predicting which amendments are likely to pass and how these might affect the final vote. Legislators must decide whether to support an amendment not just based on their preferences but also considering the potential responses from other members and the impact on the bill's overall chances of passing.

International Relations and Diplomacy

In international relations, game theory is crucial for analyzing interactions between states, particularly in conflict resolution, treaty negotiations, and strategic alliances.

Nuclear Deterrence - A classic example of game theory in international relations is the theory of nuclear deterrence. Countries with nuclear weapons must carefully calculate the consequences of various actions, including proliferation, disarmament, and potential first strikes. The mutually assured destruction (MAD) doctrine, which posits that a nuclear war would be devastating for all, acts as a deterrent, stabilizing relations between nuclear powers under a tense equilibrium.

Trade Negotiations - In trade negotiations, countries strategize to maximize their economic benefits while protecting domestic industries. Game theory models these negotiations as a series of moves and counter-moves, where each country must anticipate the concessions and demands of others, aiming to reach an agreement that is better than any available alternative.

Political Campaigns and Media Strategy

The interaction between media, public opinion, and political strategy is also a rich field for applying game theory. Politicians and parties must decide how to present their messages, react to news, and respond to opponents' statements, all within the highly strategic context of media influence.

Managing Public Opinion - In an era of rapid information flow, controlling the narrative can be pivotal. Politicians use game theory to time their announcements and to frame their messages in ways that maximize positive coverage and minimize negative reactions. For example, releasing certain news to overshadow unfavorable reports or scheduling announcements to capture peak public attention involves strategic decision-making akin to game-playing.

Public Policy and Regulation

Game theory is not just about winning elections or forming governments; it also shapes the creation and implementation of public policy.

Regulatory Capture - Consider the interaction between regulatory agencies and the industries they regulate. There's often a risk of regulatory capture, where regulators act in the industry's interest rather than the public's. Game theory helps in designing mechanisms and institutions that minimize this risk by aligning the incentives of regulators with those of the public, such as through increased transparency, better oversight, and stricter penalties for collusion.

Section 15: Military Strategy and Game Theory

Strategic Deterrence and Defense

One of the most critical applications of game theory in military strategy is in the area of deterrence, discussed also in the previous section. Deterrence aims to prevent aggression by promising a significant retaliation. Game theory helps military strategists determine the level of threat that is likely to deter potential aggressors without escalating tensions unnecessarily.

Nuclear Deterrence - Consider the delicate balance of power in nuclear deterrence strategies, where the decision to maintain or deploy nuclear weapons involves calculating potential responses from adversaries. Game theory models such scenarios as a high-stakes game where each side must consider the others' possible moves and countermoves, aiming to maintain a stable balance that discourages nuclear aggression due to the mutually assured destruction.

Conventional Deterrence - In conventional military contexts, deterrence might involve showing enough strength to dissuade enemies from attacking. Game theory helps in allocating resources across different military bases and border areas to maximize the perception of strength and readiness, influencing enemy perceptions and decisions.

War Gaming and Simulation

Military organizations frequently use war gaming and simulations to predict outcomes of battles and larger conflicts. These exercises are deeply embedded with game theory, which helps in creating realistic scenarios that test strategic decisions and their outcomes.

Operational War Games - Using game theory, military strategists can simulate engagements between their own forces and those of a potential adversary. For example, in a simulated naval conflict, game theory can help predict whether an aggressive maneuver by one side might lead the other to retreat or escalate the conflict, allowing strategists to plan accordingly.

Resource Allocation and Logistics

Effective logistics are vital in military operations, and game theory plays a key role in optimizing these processes. Decision-makers must determine how to distribute limited resources such as troops, equipment, and supplies to maximize operational effectiveness.

Dynamic Resource Allocation - Imagine the challenge of distributing resources in an ongoing conflict. Game theory models can help military logistics officers decide how to best allocate resources dynamically, responding to changing conditions on the ground. This might involve deciding which units to resupply first or how to prioritize medical and evacuation services in a way that saves the most lives and maintains combat effectiveness.

Cyber Warfare and Information Security

In the modern military landscape, cyber warfare has become a front line. Game theory is crucial in designing strategies for both attacking and defending in the cyber domain.

Defensive Strategies - In cybersecurity, game theory helps in understanding the potential moves of hackers or state-sponsored cyber attackers and in developing robust defense mechanisms. For example, a military unit might use game theory to determine the optimal way to protect sensitive information across networks, anticipating where attacks are most likely and disguising vulnerabilities.

Offensive Cyber Operations - Conversely, when planning offensive cyber operations, strategists use game theory to choose targets and timing, maximizing disruption to the enemy while minimizing the risk of severe retaliation.

Psychological Operations and Propaganda

Psychological operations (PSYOPS) aim to influence enemy and civilian attitudes and behaviors during conflicts. Game theory models the interactions between messaging (propaganda) and behavior, helping strategists plan operations that effectively alter perceptions and degrade enemy morale.

Influence Campaigns - For instance, in a scenario where military forces aim to weaken the enemy's resolve, game theory can guide the timing and content of propaganda to exploit known psychological vulnerabilities, maximizing impact and sowing confusion and discord among adversary ranks.

Alliance Formation and Coalition Warfare

In multi-nation conflicts or peacekeeping missions, forming and maintaining alliances is complex and fraught with challenges. Game theory helps in understanding the dynamics of coalition forces, each with its own goals and risk tolerances.

Strategic Alliances - By applying game theory, military strategists can better negotiate the terms of alliances, predict the behavior of allies, and plan joint operations that consider the diverse objectives and capabilities of each member. This ensures that collective action is cohesive and effective, despite differing national interests.

Section 16: Computer Science and Networking

Network Design and Traffic Management

Networks, whether they involve data centers, communication infrastructures, or the internet, require efficient management of traffic and resources to prevent congestion and ensure optimal performance.

Congestion Games in Network Routing - Imagine you are managing a network where each router must decide which path data should take to minimize latency. Game theory models these decisions as a non-cooperative game where each router independently chooses a path that minimizes its own traffic load without regard to the overall network performance. This can lead to suboptimal routing and congestion. By implementing cooperative game theory models, you can design incentives or penalties that encourage routers to choose paths that benefit the entire network, not just individual routers.

Algorithm Design and Optimization

In computer science, algorithms must often make decisions with incomplete information or compete for resources, scenarios where game theory provides crucial insights.

Algorithmic Mechanism Design - Consider an online auction platform where users submit bids within a certain timeframe, and the highest bidder wins. Designing these algorithms involves ensuring that they are not only efficient in matching offers and bids but also resistant to manipulation. Game theory helps in constructing mechanisms that incentivize users to bid their true valuations, enhancing the platform's fairness and efficiency.

Security Protocols and Cyber Defense

Cybersecurity is a critical area where game theory models the interactions between defenders and attackers, helping to anticipate breaches and design robust defense mechanisms.

Modeling Cyber Threats - In cybersecurity, you might model potential interactions with hackers as a game where the defender must allocate resources across various points of potential vulnerability. Game theory can predict the most likely targets of attackers and suggest optimal ways to distribute defensive resources, balancing the costs of defense with the need for protection.

Distributed Systems and Cooperative Computing

In distributed computing, multiple computers work together to solve complex problems, necessitating algorithms that efficiently manage the division of labor and data sharing.

Resource Allocation in Distributed Networks - Imagine a scenario where a distributed system must allocate limited computational resources to a number of tasks, each with different priorities and resource requirements. Game theory can be used to create a fair allocation system that considers the utility of completing each task, potentially prioritizing tasks that are critical or that maximize the overall benefit to the system.

Machine Learning and Data Mining

Game theory intersects with machine learning in areas such as learning in multi-agent environments, where multiple algorithms learn and adapt simultaneously.

Adversarial Machine Learning - In adversarial machine learning, algorithms are trained to handle input data that may be intentionally misleading or harmful, designed by attackers aiming to confuse the model. Game theory helps in designing these algorithms so that they can anticipate and counter adversarial strategies, improving their robustness and accuracy.

Networking Strategies in Social Media and E-commerce

In social media and e-commerce, networking strategies determine how information, products, and services are recommended and disseminated among users.

Influence and Information Cascades - Game theory models how information or behaviors spread in networks, which can be crucial for viral marketing or controlling misinformation. By understanding the strategic formation of network ties and how they influence the spread of information, platforms can enhance positive engagement and mitigate the risks associated with rapid information spread.

Peer-to-Peer Systems and File Sharing

Peer-to-peer (P2P) networks, where users share resources such as files or processing power without central coordination, rely heavily on game theory to encourage cooperative behavior among inherently selfish agents.

Encouraging Cooperation in P2P Networks - In a P2P network, game theory can design incentive schemes that encourage users to contribute resources rather than merely consume them. For example, a file-sharing network might use a tit-for-tat strategy, where users gain access to resources based on how much they share with others, promoting a balanced and efficient network.

Section 17: Exploring Game Theory in Evolutionary Biology

Survival Strategies and Reproductive Success

One of the primary applications of game theory in evolutionary biology is in understanding how organisms develop survival and reproductive strategies that maximize their fitness—often defined as the ability to survive and reproduce.

The Hawk-Dove Game - Consider the Hawk-Dove game discussed earlier, a classic model in evolutionary game theory. This model explores how aggression and cooperation can evolve in a population competing for shared resources. Hawks are aggressive and fight for the resource, risking injury, while Doves avoid conflict. Depending on the cost of fighting and the value of the resource, a stable strategy may evolve. For instance, if the cost of injury is high, Doves may fare better, promoting a more cooperative behavior within the population.

Predator-Prey Dynamics

Game theory models the complex interactions between predators and prey, providing insights into how these species evolve strategies that balance energy expenditure, risk of predation, and the need to feed.

Evolutionarily Stable Strategies in Predation - In a system where predators must decide how much time and energy to invest in hunting prey that have varying degrees of difficulty to catch, game theory helps predict what strategies might become prevalent. For example, a predator species might evolve to specialize in catching a certain type of prey if that strategy yields enough food without excessive energy expenditure or risk, thus becoming an evolutionarily stable strategy (ESS).

Sexual Selection and Mating Behaviors

Sexual selection is another area where game theory is extensively applied. The strategies that individuals employ to attract mates can have significant implications for the traits that get passed on to future generations.

The Game of Mate Choice - In many species, individuals must choose whether to invest in traits that are more likely to attract mates, such as elaborate plumage or impressive displays. Game theory models these decisions by considering the benefits of attracting mates against the costs of developing and maintaining these traits. For example, in peacock species, males with larger and more colorful tails may attract more mates but are also more visible to predators.

Cooperation, Altruism, and Social Behavior

Game theory is pivotal in exploring the evolution of cooperation and altruism among organisms, particularly in social species where individuals interact closely with each other.

The Prisoner's Dilemma and Altruism - This famous game theory model shows how cooperation may not always be the natural choice for self-interested individuals because betraying a partner can sometimes offer a higher reward. However, repeated interactions (iterated Prisoner's Dilemma) can lead to the evolution of cooperative behavior if individuals realize that long-term benefits of mutual cooperation outweigh the short-term gains from defection. This can explain why behaviors like altruism, which seem to reduce an individual's fitness in favor of another's, can persist in populations.

Kin Selection and Inclusive Fitness

Kin selection theory, which is a cornerstone of evolutionary biology, can be explored through game theory by examining how genes promoting altruistic behaviors toward relatives can spread through populations.

Hamilton's Rule and Game Theory - Hamilton's Rule states that altruism is favored by natural selection if the cost of the altruist is less than the benefit to the recipient multiplied by the degree of relatedness between them. Game theory models these interactions by quantifying costs and benefits in strategic terms, helping to predict when altruistic behavior will occur in a population.

Symbiosis and Mutualism

Game theory also applies to symbiotic and mutualistic relationships, where different species live together and benefit from each other's presence.

Modeling Symbiotic Relationships - In scenarios where different species interact closely, game theory can help explain how these interactions evolve to become mutually beneficial. For example, in the relationship between flowering plants and their pollinators, both parties have evolved strategies that maximize their respective payoffs—nutrition for the pollinators and reproductive success for the plants.

Section 18: Corporate Strategy and Negotiations

In corporate strategy and negotiations, game theory is an invaluable asset, providing tools to navigate complex strategic landscapes, make informed decisions, and negotiate effectively. By understanding and applying the principles of game theory, you can enhance your strategic capabilities, leading to better outcomes in negotiations, more effective competitive positioning, and smarter internal decision-making. Remember that the real power of this approach lies in its ability to provide a clear framework for understanding the interdependencies and potential outcomes of various corporate actions.

Strategic Decision Making in Corporate Environments

Corporations operate in competitive environments where every strategic decision can significantly impact their market position. Game theory helps in these scenarios by providing a framework to anticipate competitor reactions, enabling companies to strategize effectively.

Market Entry and Expansion - When a company considers entering a new market or expanding its product line, it faces potential retaliation or competitive moves from existing players. Game theory models such scenarios, allowing you to analyze whether entering the market will provoke aggressive price wars or if there is a cooperative strategy that might lead to mutual benefit for all players involved.

Contract Negotiations and Deal-Making

Negotiations are central to corporate strategy, whether they involve mergers, acquisitions, partnerships, or supplier agreements. Game theory provides insights into how to negotiate terms that maximize a company's benefits while ensuring the deal remains attractive to the other side.

Optimal Negotiation Tactics - For instance, in a merger negotiation, game theory can help each side understand the other's likely responses to various offers, guiding them to structure proposals that are likely to be accepted and that maximize their respective returns. Understanding the payoff matrix of different negotiation outcomes can lead to more informed and strategic decision-making.

Resource Allocation within Corporations

Effective internal resource allocation can significantly impact a company's efficiency and profitability. Game theory applies to scenarios where resources must be allocated across different departments or projects with competing needs.

Balancing Competing Interests - Imagine you're the CFO of a large corporation deciding how to allocate budget across projects that promise varying degrees of return and risk. Game theory can help in modeling the strategic moves of different departments as they compete for resources, aiding in achieving a balance that maximizes the overall company's welfare.

Competitive Strategy and Market Positioning

In competitive markets, companies must continually assess their strategies in light of their competitors' actions. Game theory aids in understanding competitive dynamics and in developing strategies that effectively counteract those of rivals.

Price Wars and Collusive Equilibria - Consider a scenario where your company and a competitor are tempted to undercut each other's pricing to gain market share. Game theory can model this competitive interaction and show that such a strategy might lead to a destructive price war, suggesting instead that maintaining stable prices could be beneficial for both players in the long run.

Managing Corporate Mergers and Acquisitions

Mergers and acquisitions are critical growth strategies for companies but involve complex negotiations and strategic planning. Game theory helps by modeling the possible outcomes of different merger strategies, considering the responses of other market players, regulators, and even the reaction of the stock market.

Strategies for Successful Mergers - When two companies consider merging, game theory analyses can predict how different merger configurations might affect the market. It can also guide the negotiating process by helping each side understand the other's pressures and constraints, leading to a mutually beneficial arrangement.

Crisis Management and Strategic Responses

In times of corporate crises—be it due to financial downturns, public relations issues, or legal troubles—game theory provides strategies for managing the situation and mitigating damage.

Game Theory in Crisis Negotiations - During a crisis, the decisions made by your company will likely provoke reactions from stakeholders, including investors, customers, and the media. Game theory can help predict these reactions and guide your strategy to manage the crisis effectively, ensuring that actions taken minimize long-term damage and align with recovery strategies.

Ethics and Corporate Responsibility

As corporations increasingly recognize their roles in societal and environmental sustainability, game theory can assist in navigating these complex areas, particularly in balancing profit motives with ethical considerations.

Sustainable Practices and Long-term Payoffs - In decisions involving environmental sustainability, game theory helps companies evaluate the long-term benefits of adopting greener practices against the short-term costs. It can model how such strategies might affect competitors' actions and consumer preferences, aiding in the development of strategies that not only benefit the company but also contribute to broader societal goals.

Section 19: Game Theory in Business

Strategic Positioning and Brand Differentiation

In the crowded marketplace, companies must strategically position their products and differentiate their brands to capture consumer attention and loyalty.

Competitive Positioning - Imagine you are a marketing director for a beverage company. Game theory helps you analyze how introducing a new product might affect the market share of your competitors and predict their potential responses, such as launching their own new products or adjusting prices. This strategic analysis supports decision-making regarding product features, market positioning, and target demographics to maximize market impact and profitability.

Advertising Wars and Media Spending

Advertising plays a key role in how brands communicate with their markets. Game theory explores the decisions companies make about advertising spending and the types of media they use.

Allocating Advertising Budgets - When two or more companies compete for market share, each must decide how much to spend on advertising and which channels to use. Game theory can model this as a game where each player's decision affects the others' outcomes, helping you optimize your advertising spend across various channels to achieve the best return on investment.

Consumer Behavior and Loyalty Programs

Understanding and influencing consumer behavior are central to effective marketing. Game theory provides insights into how consumers might respond to different marketing tactics, including loyalty programs and promotions.

Designing Effective Loyalty Programs - Using game theory, marketers can design loyalty programs that maximize customer retention and increase lifetime value. For example, by analyzing how consumers value different rewards, marketers can structure a program that motivates continued purchases and effectively differentiates from competitors' programs.

Digital Marketing and Social Media Strategies

In the digital age, social media and online platforms have become battlegrounds for brand visibility and consumer engagement. Game theory helps in crafting strategies that maximize online engagement and convert followers into customers.

Viral Marketing Campaigns - Game theory is used to plan viral marketing campaigns by modeling how information spreads through social networks. Understanding the network structures and the influence of key nodes (influential social media users) can help in crafting messages that are more likely to be shared widely, thus maximizing the campaign's reach and impact.

Market Research and Consumer Insights

Understanding market trends and consumer preferences is crucial for tailoring marketing strategies. Game theory enhances market research efforts by predicting how changes in consumer preferences might evolve under different scenarios.

Scenario Analysis in Market Trends - By using game theory, marketers can perform scenario analyses to predict how introducing a new product feature might change consumer preferences and affect competitor actions. This helps in anticipating market shifts and preparing strategies that leverage these changes effectively.

Final Reflection Questions

1) Ethical Considerations: Reflect on the ethical implications of using game theory in decision-making. Are there situations where employing game theory might be considered manipulative or unfair? How should these concerns be addressed?

2) Game Theory and Daily Life: Identify an aspect of your daily life significantly affected by the strategic decisions of others. How can understanding game theory change your approach to these situations?

3) Technological Impact: How might advances in technology, such as artificial intelligence and machine learning, impact the development and application of game theory? What new types of games might emerge?

4) Cross-Disciplinary Benefits: Consider a field of study or a hobby you are interested in. How could insights from game theory enhance your understanding or performance in this area?

5) Future of Game Theory: What do you think is the future of game theory? Which emerging trends or upcoming challenges could shape its evolution?

6) Personal Strategy Development: How has learning about game theory influenced your thinking about your own decision-making processes? Are there specific strategies you plan to apply more rigorously?

7) Contrasting Theories: Compare and contrast game theory with another decision-making framework you know (like behavioral economics, decision theory, or another relevant theory). What unique insights does game theory provide that the other framework does not?

8) Innovative Thinking: Imagine you are tasked with solving a complex problem in your community or workplace using game theory. Outline your approach and consider which game theoretic concepts would be most beneficial.

9) Critical Assessment: Reflect on the limitations of game theory. In what ways might relying on game theoretic models lead to misjudgments or errors in complex real-world situations? How can these risks be mitigated?

10) Integration with Other Disciplines: How can game theory be integrated with psychological insights to better predict and understand human behavior in competitive and cooperative scenarios? Consider how psychological factors like trust, fear, and motivation might alter theoretical predictions.

11) Global Implications: Reflect on a global issue such as climate change, international trade, or geopolitical conflicts. How can game theory be applied to understand and potentially resolve these complex problems? What game theoretic strategies might be effective, and what limitations could they face in such a broad and diverse context?

Section 20: Exploring Game Theory in Behavioral Economics

Behavioral economics integrates insights from psychology into economic theory to understand decision-making better. Game theory within this field explores how real people make decisions in strategic situations, often deviating from the perfectly rational behavior predicted by traditional economic models.

Game theory in behavioral economics provides a robust framework for understanding the complex and often irrational behaviors that underpin human decision-making. By integrating psychological insights into economic models, behavioral game theory enhances our ability to predict and influence behavior in economic settings, from individual financial decisions to broad market and policy environments. As we continue to explore these interactions, we gain not only a better understanding of economic and social phenomena but also the tools to improve welfare and efficiency across various domains.

Understanding Irrational Behaviors

Behavioral economics challenges the assumption of rational decision-making by introducing concepts such as bounded rationality, where decision-making is limited by the information available, cognitive biases, and the time available to make decisions.

Irrationality in Financial Decisions - Consider the ways individuals invest or save money. Traditional economic theory suggests that people will always make decisions that maximize their utility based on stable preferences and rational analysis. However, behavioral game theory shows that people often make decisions based on heuristic-driven biases such as overconfidence or loss aversion, leading to suboptimal financial behaviors like insufficient saving or excessive risk-taking in investments.

The Role of Social Preferences

Game theory in behavioral economics also examines how social preferences—concerns for fairness, altruism, or spite—affect economic decisions. These preferences can dramatically influence outcomes in ways that traditional economic models cannot predict.

Fairness in Wage Negotiations - For instance, in wage negotiations, game theory combined with behavioral insights can explain why some employees might reject a wage offer they see as unfairly low, even if accepting the offer is better than their next best alternative. This behavior reflects a preference for fairness and a willingness to sacrifice personal gain to prevent perceived inequities.

Behavioral Game Theory in Market Dynamics

Market behavior often deviates from the rational model in significant ways due to the bounded rationality of consumers. Behavioral game theory helps to analyze and predict market outcomes under more realistic assumptions.

Consumer Choice and Pricing Strategies - In markets, companies often use pricing strategies that exploit consumer biases. For example, decoy pricing can influence consumer choices—a higher-priced item can make a slightly less expensive item look more attractive, even if the lower-priced item is what the consumer originally intended to buy. Game theory models these interactions and predicts both consumer and competitor responses to different pricing strategies.

Behavioral Game Theory in Public Policy

Behavioral game theory has profound implications for public policy, particularly in designing interventions that correct market failures due to irrational behaviors.

Policy Design Using Nudges - Consider the use of nudges—subtle policy shifts that encourage people to make decisions that improve their welfare without restricting freedom of choice. For example, many people do not save enough for retirement, either because they underestimate how much they need, they procrastinate, or they find the process of setting up retirement savings accounts too complex. To address this, a government or employer might implement a "default nudge" by automatically enrolling employees in a retirement savings plan, such as a 401(k) in the United States, with a default contribution rate.

Negotiations and Strategic Interactions

In negotiations, understanding the psychological and strategic motivations of others can lead to better outcomes than strategies that assume rational self-interest alone.

Negotiations with Incomplete Information - Game theory in behavioral economics models negotiations where parties may irrationally overvalue their positions due to factors like the endowment effect, where individuals value what they own more highly simply because they own it. This understanding can lead to more effective negotiation tactics that consider emotional factors as well as logical arguments.

Cooperation and Competition

Exploring the dynamics of cooperation and competition through a behavioral game theory lens reveals that human cooperative behavior is often maintained by social norms and the anticipation of future interactions, rather than immediate rational calculations.

The Evolution of Trust and Cooperation - Game theory can help explain how mechanisms like repeated interactions, reputation effects, and reciprocal behavior support cooperation in business and social settings, even when short-term incentives to cheat or defect might exist.

Behavioral Insights in Contract Theory and Mechanism Design

Behavioral game theory is also essential in contract theory and mechanism design, where understanding actual human behavior leads to better-designed contracts and mechanisms that account for things like incentive misalignment and information asymmetries.

Contracts That Account for Loss Aversion - In employment contracts, understanding that employees might have a strong aversion to wage cuts—even more than they would appreciate equivalent wage increases—can lead to better-designed compensation packages that minimize dissatisfaction and turnover.

Section 21: Game Theory Applications in Finance

Portfolio Management and Investment Strategies

In world of investment, portfolio managers and individual investors use game theory to optimize their investment strategies, considering the potential moves and reactions of other market participants.

Balancing Risk and Return- Imagine you are an investor deciding how to allocate your portfolio among stocks, bonds, and other assets. Game theory helps you understand the potential moves of other investors in response to economic changes, allowing you to adjust your strategy to maximize returns while managing risk. For instance, if you anticipate a market downturn based on game-theoretic analysis of investor behavior, you might increase your holdings in safer assets like bonds.

Market Competition and Trading

In trading, participants must constantly anticipate the actions of others, whether they are competing to buy the same securities or trying to sell before a market downturn. Game theory models these interactions and helps traders develop strategies that anticipate and capitalize on market movements.

High-Frequency Trading (HFT) - High-frequency trading is an area where game theory is particularly prevalent. Traders use algorithms that can execute trades in milliseconds, often based on predictions of how other traders will react to market news or changes. Game theory helps in designing these algorithms to outperform human traders and other algorithms by predicting their likely actions.

Corporate Finance Decisions

Corporations often face strategic financial decisions that involve interactions with other stakeholders, including investors, regulators, and other firms. Game theory is crucial in making decisions about issues such as financing, dividend policies, and mergers and acquisitions.

Negotiating Mergers and Acquisitions - When a company considers a merger or acquisition, it must not only assess the financial implications but also predict how different stakeholders will react. For example, if acquiring a particular company might lead to antitrust concerns, game theory can help predict how regulators might respond and whether competitors might try to block the merger.

Risk Management

Risk management involves identifying, assessing, and prioritizing risks followed by coordinated efforts to minimize, monitor, and control the probability or impact of unfortunate events. Game theory provides a framework for understanding the interdependent risks that different financial decisions create.

Insurance and Hedging Strategies - In the insurance industry, companies use game theory to price policies based on the probability of events and the expected behavior of policyholders. Similarly, in hedging, financial institutions use game theory to construct portfolios that minimize the risk of adverse price movements in an asset.

Regulatory Compliance and Financial Policy

Financial markets are heavily regulated to prevent fraud, ensure fair trading, and protect investors. Game theory is used to anticipate how companies might respond to various regulatory strategies and to design regulations that effectively deter undesirable behavior without stifling economic activity.

Compliance Strategies - Financial regulators use game theory to determine the optimal level and type of enforcement that will encourage compliance with financial regulations. For example, they might use game theory to balance the cost of regulatory compliance against the benefits in terms of reduced risk of financial crises.

Behavioral Finance

Behavioral finance, which looks at the influence of psychology on the behavior of investors and the subsequent effect on markets, also uses game theory to analyze how irrational behaviors affect financial decisions.

Market Bubbles and Crashes - Game theory can help explain how psychological factors contribute to market bubbles and crashes. For instance, during a bubble, game theory can model the herd behavior of investors who buy overvalued assets because they see others doing the same, expecting to sell the assets at an even higher price.

Strategic Financial Negotiations

Financial negotiations, whether related to restructuring debt, negotiating terms of a financial bailout, or setting up strategic partnerships, can be complex, with multiple parties and competing interests.

Debt Restructuring Negotiations - In debt restructuring, game theory helps creditors and debtors predict and influence each other's strategies. Creditors need to balance the desire to recover as much as possible with the need to keep the debtor solvent, while debtors must negotiate terms that allow them to sustain operations and return to profitability.

Section 22: Game Theory in Law and Regulation

Game theory offers valuable insights also into law and regulation by modeling the strategic behavior of individuals, corporations, and governments within legal frameworks. Understanding these interactions helps policymakers design more effective laws and helps participants navigate the legal environment more successfully.

Strategic Litigation and Legal Bargaining

In the legal domain, much of the interaction between parties can be seen as strategic games where each party seeks to maximize their outcomes, whether through settlements, trials, or arbitration.

Settlement Negotiations - Consider the dynamics of a settlement negotiation in a civil lawsuit. Plaintiffs and defendants assess the strengths and weaknesses of their cases, predict the likely outcome of going to trial, and use this information to negotiate a settlement. Game theory models these interactions by assessing the risks and payoffs associated with various strategies, such as settling early versus pursuing a lengthy court battle. It helps parties reach a mutually agreeable settlement by evaluating the expected utilities of different outcomes.

Compliance and Regulatory Enforcement

Game theory is crucial in understanding how entities respond to regulations and the enforcement actions of regulatory bodies. It helps in designing enforcement mechanisms that effectively deter violations while promoting compliance.

Corporate Compliance Programs - For instance, a corporation deciding how much to invest in a compliance program must weigh the cost of the program against the potential costs of non-compliance, including fines, reputational damage, and other penalties. Game theory helps corporations strategize their compliance efforts by predicting regulatory behaviors and the likelihood of inspections or audits, ensuring that compliance is both effective and efficient.

Antitrust and Competitive Strategy

In antitrust law, regulators use game theory to analyze the competitive effects of mergers, acquisitions, and other business practices. The strategic decisions of firms in competitive markets can significantly impact consumer welfare and market structure.

Mergers and Acquisitions Analysis - Game theory models the strategic interactions between firms in a merger scenario to predict whether the merged entity will have the incentive and ability to raise prices or stifle competition. This analysis helps antitrust authorities decide whether to approve, reject, or require modifications to merger proposals to maintain competitive markets.

Environmental Regulation and Public Goods

Game theory also applies to environmental law and the regulation of public goods, where individual incentives to exploit resources must be balanced against societal needs for sustainability.

Pollution Credits and Cap-and-Trade Systems - In environmental regulation, game theory is used to design cap-and-trade systems for pollution control. Companies receive or buy pollution credits, and they must decide whether to sell these credits or use them to cover their emissions. Game theory helps in predicting how companies will respond to prices of pollution credits and the total cap on emissions, optimizing environmental outcomes while allowing economic activities.

Intellectual Property and Innovation

The strategic interaction between firms regarding intellectual property (IP) rights can significantly influence innovation and economic growth. Game theory provides insights into how firms might behave in scenarios involving patent races, licensing agreements, or litigation.

Patent Races and R&D Investment - In industries where technology evolves rapidly, companies often race to patent new discoveries. Game theory models these races, helping firms decide how much to invest in research and development (R&D), considering the potential payoffs of securing a patent and the competitive risks of falling behind.

Criminal Law and Deterrence

Game theory in criminal law relates to the deterrence effect of legal penalties on potential lawbreakers. It analyzes how different strategies of law enforcement and penalty setting can effectively reduce crime.

Optimal Penalties for Deterrence - Authorities must decide the severity of penalties for various crimes in a way that deters wrongdoing without being overly harsh. Game theory helps in determining the optimal level of penalties that maximize deterrence based on the likelihood of apprehending lawbreakers and the societal costs of different crimes.

Section 23: Game Theory in Social Network Analysis

Understanding Influence and Power Dynamics

In any social network, some individuals or nodes hold more influence than others. This influence can be due to their position within the network, their connections to other influential nodes, or their access to critical resources or information.

Centrality and Power - Imagine you're analyzing a corporate network to determine which employees are most influential in spreading information. Using game theory, you can model scenarios to see how information from one central node might spread differently compared to information from a more peripheral node. Game theory helps predict which nodes are most crucial for ensuring efficient information dissemination or for controlling the spread of information in the network.

Coalition Formation in Social Networks

Coalitions or cliques within networks often form when groups of nodes (individuals or organizations) align themselves to achieve common objectives, which might be social, economic, or political.

Strategies for Forming Alliances - In the context of political networks, game theory can help explain why certain alliances form during elections or legislative processes. These models consider the benefits each member gains from the alliance against the costs, such as compromises they must make on their positions. By understanding these dynamics, you can predict how stable an alliance might be or what factors could lead to its dissolution.

Information Diffusion and Viral Marketing

Game theory is particularly useful in analyzing how information or trends spread in social networks, which is crucial for viral marketing campaigns or public health information dissemination.

Modeling Information Spread - If a company wants to launch a new product using a viral marketing strategy, game theory can model the potential 'paths' the information might take through social networks. It helps marketers understand which nodes (influential users) should be targeted to maximize the spread and impact of their campaign based on the users' positions and connections within the network.

Reciprocity and Social Exchange

In social networks, interactions often involve reciprocity where actions by one individual lead to a reciprocal action by others. Game theory models these interactions to understand and predict cooperative behaviors.

Exchange Dynamics - Consider a professional network where members share job opportunities or business advice. Game theory can analyze how reciprocity influences the willingness of members to contribute valuable information. It models scenarios to determine when members might withhold information, fearing over-exploitation, or when they might share freely, expecting future reciprocation.

Negotiation and Conflict Resolution

Social networks often witness conflicts or negotiations, whether in online communities, among businesses in a cluster, or in international relations. Game theory provides a framework for analyzing these scenarios to find potential resolutions.

Negotiation in Online Platforms - In an online community, for example, game theory can help community managers design mechanisms that encourage users to resolve disputes amicably. It can predict which negotiation strategies might lead to successful conflict resolution based on past interactions and the relative influence of parties involved.

Behavioral Insights in Network Interactions

Game theory combined with behavioral economics gives deeper insights into the non-rational behaviors exhibited by network participants, influenced by biases, heuristics, and social preferences.

Understanding Non-Rational Behaviors - In social networks, people do not always act in purely self-interested ways; they are often influenced by perceived social norms, desires to fit in, or react against perceived injustices. Game theory helps in modeling these behaviors to understand how they impact network dynamics, such as the spread of social movements or the enforcement of social norms.

Section 24: Game Theory in Public Policy and Resource Allocation

Managing Public Goods and Common Resources

Public goods and common resources such as parks, clean air, and water bodies pose significant challenges in management due to their non-excludable and non-rivalrous nature, which often leads to overuse or under-provision.

Tragedy of the Commons - Consider the scenario of local fishermen accessing a shared fishery, which is susceptible to overfishing. Game theory models this situation as a common resource game where each fisherman decides how much to fish, knowing that overfishing could deplete the resource but also facing the immediate benefit of a larger catch. Through game theory, policymakers can devise quota systems or cooperative agreements that align individual fishermen's incentives with sustainable fishing practices.

Budget Allocation and Public Spending

In government, deciding how to allocate a limited budget across various public sectors such as education, healthcare, and defense involves strategic decision-making where different groups lobby for more resources.

Optimal Budgeting Decisions - Game theory helps in modeling these lobbying interactions and can guide the allocation process to achieve maximum social welfare. For example, it can be used to predict the outcomes of increasing education funding on future economic growth versus allocating more to immediate healthcare needs, helping policymakers strike a balance based on the predicted responses of affected groups.

Social Welfare Programs and Incentive Structures

Designing social welfare programs such as unemployment benefits, food subsidies, and housing vouchers requires careful consideration of the incentives they create and their impacts on behavior.

Designing Effective Incentive Schemes - Game theory is instrumental in predicting how individuals will respond to different structures of welfare benefits. For instance, it can analyze whether higher unemployment benefits discourage job-seeking or if they provide necessary support that increases the quality of job matching. Policymakers can use these insights to design programs that encourage desirable behaviors while providing needed support.

Educational Policies and Access

Game theory also informs educational policy, particularly in areas such as school choice, funding formulas, and access to higher education.

School Choice Systems - Using game theory, policymakers can design school choice mechanisms that allow parents to select schools for their children in a way that balances parental preferences with considerations for equity and educational effectiveness. These models help ensure that school choice systems do not inadvertently lead to increased segregation or reduced quality in certain schools.

Section 25: Game Theory in Environmental Strategy

Conservation Efforts and Wildlife Management

Conservation strategies often involve multiple stakeholders with competing interests. Game theory helps in understanding these dynamics and designing policies that encourage cooperation for the preservation of ecosystems and wildlife.

Protecting Endangered Species - Imagine a scenario where multiple landowners control areas that are crucial for the migration of an endangered species. Each landowner faces the choice of developing their land for immediate economic gain or conserving it to support the species. Game theory models these decisions, showing how conservation agreements and incentives can align individual landowner interests with broader ecological goals.

Sustainable Resource Use and Management

Sustainable management of natural resources like forests, water bodies, and fisheries requires careful coordination to avoid the tragedy of the commons, where individual users acting in their self-interest deplete or degrade common resources.

Fisheries Management - In fisheries, game theory is used to model the decisions of individual fishers and to design quota systems that prevent overfishing. By understanding how fishers respond to different management policies, regulators can set catch limits that ensure the long-term sustainability of the fishery while still providing economic benefits to the communities that depend on it.

Climate Change Mitigation and Adaptation

Addressing climate change involves global coordination and negotiation, with game theory providing insights into the strategic behaviors of different nations as they negotiate international agreements.

International Climate Agreements - Consider the negotiations surrounding international climate agreements like the Paris Accord. Game theory helps understand the incentives of different countries to participate or defect based on their economic interests, environmental vulnerability, and the actions of other nations. It can guide the design of agreements that make cooperation attractive and stable by ensuring that all parties see clear benefits in compliance.

Energy Production and Consumption

The transition to sustainable energy sources is fraught with challenges that include not only technological and economic factors but also the strategic interests of different stakeholders.

Renewable Energy Adoption - Game theory models the decisions of consumers, businesses, and governments in adopting renewable energy technologies. It can predict how subsidies, tariffs, and other policy measures might influence these decisions and how competitors in the energy market will react, shaping the overall strategy for energy transition.

Urban Planning and Green Infrastructure

Urban planning decisions also benefit from game theory, especially in the integration of green infrastructure like parks, green roofs, and stormwater systems that enhance urban sustainability.

Incentives for Green Infrastructure - Game theory can analyze how different stakeholders, including property owners, developers, and municipal governments, will react to incentives for green infrastructure. By understanding these reactions, cities can design incentives that effectively encourage the adoption of green technologies and practices, balancing public and private benefits.

Section 26: Game Theory in Healthcare and Epidemiology

Hospital Resource Allocation

Hospitals must often make tough decisions regarding the allocation of limited resources like ICU beds, ventilators, and medical personnel, particularly during crises such as the COVID-19 pandemic.

Optimizing ICU Bed Utilization - Using game theory, hospital administrators can model the decision-making process for ICU bed allocation, considering factors like patient health status, likelihood of recovery, and overall impact on life-saving potential. Game theory provides a structured approach to these decisions, ensuring that resources are used efficiently while being ethically justified.

Treatment Protocols and Patient Compliance

Designing treatment protocols and ensuring patient compliance are critical in healthcare delivery. Game theory analyzes how patients might respond to different treatment plans and what strategies healthcare providers can employ to enhance compliance.

Encouraging Adherence to Treatment Plans - For chronic diseases like diabetes or hypertension, patient non-compliance with treatment regimens is a major challenge. Game theory can help in designing incentive schemes or educational campaigns that motivate patients to follow prescribed treatments more faithfully, considering their preferences and constraints.

Antibiotic Resistance and Healthcare Strategies

Antibiotic resistance is a growing concern globally, driven largely by the overuse and misuse of antibiotics. Game theory helps in understanding the dynamics of antibiotic prescription practices and developing strategies to curb resistance.

Balancing Treatment Efficacy and Resistance Management - Game theory models the decision-making process of physicians who must choose whether to prescribe antibiotics based on patient symptoms and the potential for developing resistance. It helps design protocols that balance immediate patient care needs with long-term public health goals, encouraging prudent antibiotic use.

Behavioral Health Interventions

Behavioral health interventions, including programs to reduce smoking or obesity, benefit from game theory by providing insights into how individuals might respond to various intervention strategies.

Designing Effective Health Campaigns - Game theory helps public health officials design campaigns that effectively change health behaviors. By understanding the game-theoretic interactions between personal benefits, social influences, and intervention incentives, officials can create more compelling and effective health messages and programs.

Epidemiological Modeling

Epidemiological modeling often uses game theory to predict how diseases will spread within populations and to evaluate the potential impacts of public health interventions.

Predicting Disease Spread and Intervention Success - Game theory models how individuals' decisions regarding vaccination, social distancing, or seeking medical help influence disease spread. These models can predict the outcomes of different public health strategies, helping authorities choose interventions that are most likely to manage a public health crisis effectively.

Section 27: Game Theory in Sports

On-Field Tactics and Game Strategies

In every sport, coaches and players face strategic decisions that can affect the outcome of a game. Game theory helps in analyzing these decisions to optimize tactics and strategies that maximize the chances of winning.

Optimal Play Calling in Football - Consider a football coach deciding whether to run or pass on a crucial third down. Game theory can model the defensive team's likely strategies based on previous plays, helping the coach choose a play that statistically has the best chance of success given the defensive setup they anticipate.

Player Performance and Contract Negotiations

Game theory is crucial in the business side of sports, particularly in contract negotiations where players and teams must reach agreements that satisfy both parties.

Negotiating Player Contracts - Imagine a star athlete nearing the end of a contract. Game theory helps the player's agent and the team's management forecast the potential outcomes of various negotiation tactics. For instance, the agent might leverage the player's performance metrics and market demand to maximize salary offers, while the team evaluates its salary cap and roster needs to decide how far to stretch financially.

Team Formation and Roster Decisions

Building a winning team requires careful consideration of player selection and roster dynamics. Game theory provides insights into how to assemble a team that balances skills, potential synergies, and budget constraints.

Draft Strategies and Player Trades - During sports drafts or trade periods, teams decide which players to select or trade to maximize their performance throughout the season. Game theory models these decisions, considering how other teams might act, to strategize picks and trades that enhance team capabilities without overshooting budget limits.

Betting and Predictive Analysis in Sports

Betting in sports and fantasy league decisions often utilize game theory to predict outcomes and decide bets or player selections based on probabilities of various game results.

Sports Betting Strategies - In sports betting, game theory can help bettors decide how to place bets to maximize returns based on the odds and the likely outcomes of games. By analyzing how odds might shift in response to betting patterns, bettors can find value bets where the potential payout outweighs the risk.

Fan Engagement and Marketing Strategies

Sports marketing professionals use game theory to craft campaigns that maximize fan engagement and profitability. This includes ticket pricing, merchandise sales, and promotional events.

Dynamic Ticket Pricing - Game theory helps sports franchises set dynamic ticket prices that maximize revenue based on demand, which can vary according to factors like team performance, opponent strength, and game timing. Analyzing how fans are likely to respond to different pricing tiers enables franchises to adjust prices in real-time to fill stands and optimize revenue.

Sportsmanship and Ethical Considerations

Game theory also explores the ethical dimensions of sports, such as decisions that involve sportsmanship versus competitive advantage.

Fair Play vs. Strategic Fouls - In many sports, situations arise where players must decide between playing fairly or committing strategic fouls to gain an advantage (e.g., fouling in basketball to prevent a score). Game theory can analyze the potential payoffs of such decisions, helping players and coaches weigh the immediate benefits of a foul against potential long-term reputational damage or penalties.

The Key Insights and Lessons from Game Theory to Remember

1. The Power of Strategic Thinking

Game theory emphasizes the importance of strategic thinking, which involves considering not just the immediate actions, but also the broader consequences of these actions over time. This form of analysis requires forecasting the potential moves of others and adjusting one's strategy in response. Strategic thinking shaped by game theory helps individuals and organizations to prepare for various possible futures, making it possible to navigate complex situations more effectively. It teaches the value of foresight in decision-making, emphasizing the need to plan not only for likely outcomes but also for less probable scenarios that could have significant impacts.

2. Understanding Incentives and Motivations

At its core, game theory is about understanding the incentives and motivations that drive the behaviors of individuals and groups. This understanding allows for the prediction and manipulation of actions based on the perceived benefits and costs associated with them. By analyzing what motivates different stakeholders in any given situation, game theory provides insights into how to structure interactions that align others' incentives with one's own goals. This can lead to more effective negotiation, conflict resolution, and cooperation strategies.

3. The Importance of Cooperation

Game theory famously illustrates through models like the Prisoner's Dilemma that cooperation can often lead to better collective and individual outcomes than competition. It challenges the conventional wisdom that self-interest always leads to the best outcome by showing how mutual cooperation can benefit all parties involved more than if each were working solely towards their own ends. This concept has implications for everything from international politics to everyday interpersonal relationships, suggesting that building trust and collaboration can lead to more sustainable success than adversarial approaches.

4. Predicting Outcomes

One of the primary uses of game theory is in the prediction of outcomes in interactions involving multiple actors with potentially conflicting interests. By modeling these interactions, analysts can anticipate likely moves and countermoves in complex systems such as markets, political arenas, or competitive industries. This predictive capability is essential for strategic planning and helps decision-makers choose actions that are more likely to result in favorable outcomes, based on a structured understanding of the dynamics at play.

5. Managing Risk and Uncertainty

Game theory provides valuable insights into managing risk and uncertainty, particularly in situations where not all variables are known or controllable. It teaches how to make reasoned decisions in the face of incomplete information by evaluating the probabilities of various outcomes and their potential impacts. This approach is crucial in fields like finance and insurance but is also applicable to personal decision-making, where the ability to navigate uncertainty with strategic thinking can significantly affect outcomes.

6. Evolution of Strategies Over Time

Finally, game theory explores how strategies evolve over time, adapting to changing circumstances and past outcomes. This dynamic aspect of game theory reflects the real-world environments in which strategies must not only be effective in a single instance but also over time and across different contexts. It considers the feedback loops that exist in competitive environments, where strategies are continually refined in response to others' actions and the shifting landscape. This teaches the importance of flexibility and learning in strategy development, ensuring that approaches remain relevant and effective as conditions change.

Conclusion

Throughout this book, we have explored the profound impact of game theory on a variety of fields—each chapter illustrating how this powerful mathematical tool can be applied to enhance understanding and solve complex problems in real-world scenarios. From economics and corporate strategy to healthcare, sports, and beyond, game theory provides a unique lens through which we can view the strategic interactions that shape human behavior and organizational dynamics.

At its core, game theory is about choices—specifically, how individuals or entities make decisions when their outcomes depend not only on their own actions but also on the actions of others. The foundational concepts of Nash Equilibrium, strategic dominance, and evolutionary stability have been recurring themes, helping us understand how actors behave in various strategic settings, from market competitions to environmental negotiations and social interactions.

One of the primary benefits of game theory is its ability to improve strategic decision-making. By anticipating the potential responses of others, decision-makers can better plan their strategies to achieve desired outcomes. This is evident in corporate negotiations, where understanding the probable counter-moves of another company can lead to more favorable terms, and in advertising, where companies strategize to outmaneuver competitors in capturing consumer attention.

In public policy and regulation, game theory serves as a critical tool for designing mechanisms that align individual incentives with social goals. Whether through pollution credit trading schemes or healthcare protocols, game theory helps ensure that policies are both efficient and fair, promoting broader societal welfare while considering the strategic behavior of those affected by the policies.

Beyond organizational and economic contexts, game theory has significant societal implications. It aids in addressing public goods dilemmas, managing common resources, and understanding social dynamics that influence public health, legal systems, and environmental conservation. The ability of game theory to model complex interactions makes it invaluable for crafting solutions that require cooperation and coordination among multiple stakeholders.

As we look to the future, game theory continues to evolve, particularly with advancements in computational techniques and data analytics. These tools enhance our ability to solve larger, more complex game-theoretic models, providing more detailed and accurate insights. However, challenges remain, particularly in the area of behavioral game theory, where understanding the deviations from rational behavior can lead to even more effective strategies in business and policy-making.

The interdisciplinary applications of game theory underscore its integrative power. By drawing on insights from psychology, sociology, economics, and other fields, game theory enriches our understanding of strategic interactions in diverse settings. This integrative approach not only broadens the applicability of game theory but also enhances its effectiveness in solving real-world problems.

Game theory is not merely an academic curiosity; it is a vital strategic tool with extensive applications across various fields. Its ability to model complex interactions and predict outcomes makes it indispensable for anyone involved in areas where strategic decisions are critical. Understanding and applying game theory can lead to better strategies, more effective decisions, and optimal outcomes.

As you reflect on the insights provided throughout this book, consider how the principles of game theory can be applied in your own professional and personal life. Whether negotiating a contract, planning a marketing strategy, participating in public policy discussions, or engaging in personal life endeavors game theory offers tools that can enhance your ability to think strategically and act effectively in a complex, interconnected world. Let this book be an inspiration for you to explore further the rich and rewarding applications of game theory.

Thank You!

As the journey through this book concludes, we are filled with gratitude toward each reader who has chosen to spend time within these pages. Your interest in exploring the strategic complexities and the beauty of game theory is what ultimately gives this book purpose and meaning.

While *the act of sharing your insights and experiences with the book is entirely optional*, please know that it is also immensely valued. If you feel inspired to share your thoughts, there are a couple of ways you can do so:

Option 1: Create a short video review showcasing the book. Feel free to share what you found most intriguing or how this book has impacted your understanding of game theory.

Option 2: Take a picture of the book in your favorite reading spot, accompanied by a short text describing your thoughts or reflections. If you prefer, just a brief written review is also incredibly helpful.

Each piece of feedback not only enriches the experience for other readers but also contributes immensely to our ongoing journey. Your voices and perspectives are what make the discourse around game theory vibrant and ever-evolving.

Thank you once again for your time and engagement. Whether through words or images, your feedback is a cherished part of this adventure.

To leave your review scan the QR-code below:

Get Your Bonus Content

To get the bonus content scan the QR-code or go to the link below

Or go to the following link → rebrand.ly/gt-extra

References

- fastercapital.com
- Tshilidzi Marwala. "Artificial Intelligence, Game Theory and Mechanism Design in Politics", Springer Science and Business Media LLC, 2023
- Tanmoy Hazra, Kushal Anjaria, Aditi Bajpai, Akshara Kumari. "Applications of Game Theory in Deep Learning", Springer Science and Business Media LLC, 2024
- "Complex Social and Behavioral Systems", Springer Science and Business Media LLC, 2020
- www.fool.com
- "Game Theory in Smart Grids: Strategic decision-making for renewable energy integration", Sustainable Cities and Society, 2024
- Wang, Yingjie, Zhipeng Cai, Guisheng Yin, Yang Gao, Xiangrong Tong, and Guanying Wu. "An incentive mechanism with privacy protection in mobile crowdsourcing systems", Computer Networks, 2016.
- Schmidt, Christian. "Rupture versus continuity in game theory : Nash versus Von Neumann and Morgenstern", Routledge Advances in Game Theory, 2002.
- Claudia Greco, Anna Esposito, Gennaro Cordasco, Olimpia Matarazzo. "Reciprocit versus Self-Interest in a Competitiv Interaction Context: An Experimental Study", Psychological Reports, 2022
- cedelft.eu
- Ming Jiang, Jingchao Li. "When do stable matching mechanisms fail? The role of standardized tests in Chinese college admissions", China Economic Review, 202343
- Yuehai Wang. "Chapter 9 Evolutionary Game Theory Based Cooperation Algorithm in Multi-Agent System", IntechOpen, 2009
- "Science without Laws", Walter de Gruyter GmbH, 2007
- Robert H. Frank. "6. The Burden of False Beliefs", Walter de Gruyter GmbH, 2016
- Springer Series in Reliability Engineering, 2015.
- Tuyls, K.. "What evolutionary game theory tells us about multiagent learning", Artificial Intelligence, 2007
- "Learning processes among players", Cognitive Economics, 2008
- Cable, Daniel M., and Scott Shane. "A Prisoner's Dilemma Approach to Entrepreneur-Venture Capitalist Relationships", The Academy of Management Review, 1997.
- P. Vanderschraaf. "Game Theory Meets Threshold Analysis: Reappraising the Paradoxes of Anarchy and Revolution", The British Journal for the Philosophy of Science, 12/01/2008
- towardsdatascience.com
- David L. Weimer. "The Potential of Contingent Valuation for Public Administration Practice and Research", International Journal of Public Administration, 2007
- Vega, Clara, and Christian S. Miller. "Market MicrostructureMarket microstructure", Encyclopedia of Complexity and Systems Science, 2009.
- J E Riggs. "Medical ethics, logic traps, and game theory: an illustrative tale of brain death", Journal of Medical Ethics, 2004
- Journal of Manufacturing Technology Management, Volume 23, Issue 8 (2012-10-13)
- Ojo Emmanuel Ige, Festus Rotimi Ojo, Sunday Amos Onikanni. "Chapter 14 Rural and Urban Development: Pathways to Environmental Conservation and Sustainability", Springer Science and Business Media LLC, 2024
- Ifzal Ahmad, M. Rezaul Islam. "The Fabric of Society: Understanding Community Development", Emerald, 2024

- Lia Thomson, Daniel Priego Espinosa, Yaniv Brandvain, Jeremy Van Cleve. "Linked selection and the evolution of altruism in a family-structured population", Cold Spring Harbor Laboratory, 2022
- Zengchang Qin, Farhan Khawar, Tao Wan. "Collective game behavior learning with probabilistic graphical models", Neurocomputing, 2016
- www.inderscience.com
- www.investopedia.com
- "M814 unit 6 software activities WEB57020", Open University
- Binzhen Wu, Xiaohan Zhong. "Matching inequality and strategic behavior under the Boston mechanism: Evidence from China's College admissions", Games and Economic Behavior, 2020
- Bottazzi, G.. "A laboratory experiment on the minority game", Physica A: Statistical Mechanics and its Applications, 20030601
- Glenn E. Weisfeld, Peter LaFreniere. "Emotions, not just decision-making processes, are critical to an evolutionary model of human behavior", Behavioral and Brain Sciences, 2007
- IFIP Advances in Information and Communication Technology, 2013.
- Jean-Pierre P. Langlois, Catherine C. Langlois. "Dispute Settlement Design for Unequal Partners: A Game Theoretic Perspective", International Interactions, 2007
- Su Xiu Xu, Jianghong Feng, George Q. Huang, Yue Zhai, Meng Cheng. "Toward efficient waste electric vehicle battery recycling via auction-based market trading mechanisms", International Journal of Production Research, 2022
- Shoham, Y.. "On Rational Computability and Communication Complexity", Games and Economic Behavior, 200104
- Tommi Lehtonen. "Niṣkāmakarma and the Prisoner's Dilemma", Sophia, 2020
- K.M. Ariful Kabir, MD Shahidul Islam, Sabawatara Nijhum. "Exploring the performance of volatile mutations on evolutionary game dynamics in complex networks", Heliyon, 2023
- Kubilay Geçkil, . "Strategy and Game Theory Concepts", Applied Game Theory and Strategic Behavior, 2009.

Made in United States
Troutdale, OR
06/18/2024